READING THE GLASS

READING
THE GLASS

A Captain's View of Weather,
Water, and Life on Ships

Elliot Rappaport

DUTTON

DUTTON

An imprint of Penguin Random House LLC
penguinrandomhouse.com

Illustrations by Matthew Twombly

LIBRARY OF CONGRESS CATALOGING-IN-PUBLICATION DATA
has been applied for.

ISBN 9780593185056 (hardcover)
ISBN 9780593185070 (ebook)

Printed in the United States of America
1st Printing

BOOK DESIGN BY TIFFANY ESTREICHER

To Karen

CONTENTS

CONTENTS

Where is your ancient courage? You were used
To say that extremity was the trier of spirits;
That common chances common men could bear;
That when the sea was calm, all boats alike
Showed mastership in floating.

CORIOLANUS, ACT 4, SCENE I

READING THE GLASS

0

DAVIS STRAIT

1994

There is no night in arctic summer, just a diffuse cold glow that struggles up from underneath the horizon and dissolves slowly into the sky, the sea surface black and laced with mist like exhalations from a chest freezer. At 0215 the third mate, wearing enough clothes for a walk in space, comes below with an update on the weather. We were flying along like a gull when I went to bed. Now the wind, steady from the southwest for the last twenty-four hours, is getting light and starting to veer. The barometer has plummeted from 1016 to 994 millibars and seems to be reaching bottom. It suddenly feels like a good time to shorten sail. As the next watch comes on deck, we put the extra hands to work and reduce our canvas by about half—to one small headsail, the foresail, and a storm trysail.

The breeze quits with a final sigh, and the air turns sharply cold. The off-going watch finishes their work, waves an insincere farewell, and departs to their bunks. A gentle wind begins from the north. An hour later it is blowing a gale. We strike the foresail. The storm trysail blows out at its clew fitting, threshing angrily back and forth across the quarterdeck until we take it down also. By breakfast there are gusts to force 10—storm strength—driving a cold spray of water like birdshot across the open deck.

It is pointless now to think of making progress. We start the engine and run it slowly ahead to avoid being blown too far back the way we've just come. I have the helm put to windward, and between thrust from the propeller and the balancing force from our one remaining sail our little schooner sits like a merganser, rising and falling through an arc of thirty feet with the developing seas just ahead of the beam. Under a vague brightening in the sky, the water turns from black to gray, stirred into white streaks by the gale. The air temperature is 38 degrees Fahrenheit, just slightly colder than the seawater. The wind howls. It is hard to keep your eyes open and I wish for a swim mask.

The sea has built steeply with the reversal in wind direction, and tall green waves rear over the deck and slide below us as the boat heaves up and down. We sink into the trough. The next sea stands up vertically alongside us, a deep pane of aquarium glass. I look in and glimpse sudden dark shapes moving behind—a pod of pilot whales, jet-black two-ton mammals swimming toward me, right at eye level. I dive for cover. The boat rides up the crest and the animals break the surface where we'd just been, as though they are performing at SeaWorld. In unison they cut a neat arc through the air and vanish under the stern.

It is time for coffee. We have lifelines rigged around the deck, wrestling-ring style, to keep people away from the rail. I carom forward to the galley hatch and time the motion so as to fall through as gracefully as possible. Down below, it is startlingly quiet. The crew are asleep in their bunks, wedged in behind improvised pillows of spare gear. There is one person upright at the cabin table, pensively eating a Snickers bar and staring into the middle distance. Bob is our doctor, a semiretired and much beloved family physician from the small coastal town where I live in Maine. He looks a little like the actor Burt Lancaster. I've learned that he was a PT boat officer in the Pacific before medical school and has missed being at sea for most of his life since. I am under strict instructions to bring him home alive.

I find a clean mug and fill it from the thermos, pausing to wonder just how I've gotten myself into this situation. A good question. Professional sailors are a bit like commercial pilots—licensed operators in the transportation industry, with credentials built from a blend of hands-on experience and regulatory vetting. This said, seafaring in the jet age is a rarer and less visible career path, particularly for Americans. Most of my college friends were off to jobs in law, medicine, or academia. Uninterested in such things, I'd set out sailing until an avocation became a profession, and now here I was—responsible for an old wooden ship and sixteen souls as the cold rain blew sideways and the sea built into steep gray ridgelines around us.

It turns out that there is still nothing like a sailing ship to help you care about the weather. Motor vessels don't need the wind to take them where they are going, though their crews still recognize that weather at its extremes can take over their plans. For any who choose the more quixotic means of sail transport, the weather is everything. Understanding the weather means getting successfully to where you are going and knowing when not to try. At the least, knowledge adds a dimension of clarity to the miserable days spent amid conditions you can do nothing about.

Complex as the outcomes may be, the drivers of weather are built around a small core of scientific principles: The atmosphere is a mixture of gases subject to heating at different rates by solar radiation, governed by the physical laws uniting temperature and pressure. Weather is just atmospheric stirring, a three-dimensional cycle that solves imbalances by redistributing energy around the globe. This mixing is torqued by the rotational momentum of Earth and coupled with a water cycle that carries heat and moisture endlessly from place to place. Scientists have understood this qualitatively for some time, but it is really only in the modern era of data-gathering—the age of aircraft, satellites, and supercomputers—that a real-time imaging of atmospheric processes has become possible.

The plans for sailing passages are drawn with the expected wind and sea conditions set atop every page. Conditions change, and plans evolve. Each day's actions are steered by the steady uptake and processing of information. What do we see? What do the forecasters say? What happened on the last trip? The gathering of experience turns novices into experts. Sailors tell tales of weather as a pastime, but ultimately their storytelling serves the higher purpose of adding to accumulated knowledge—building onto a shared understanding of a complex and dangerous environment.

Like pilots, roofers, and mountain climbers, mariners are by default obsessed with the weather, immersed in it as part of their daily calculus. I've been a student of weather since high school, but in hindsight I recognize this particular storm as a tipping point in the process—a watershed in realizing how completely my profession was enmeshed with a system that was by its nature never quite the same. This has felt more like a relationship than a course of study—a career-long game of Rubik's Cube, where one side of the cube is your plan and the other five are built from the ever-moving parts of sea and atmosphere.

Down in the cabin I look under the floorboards to see if we are sinking. I know that someone else on the crew is doing this already, but in heavy weather the captain of every wooden ship envisions the thousand stitched-together pieces that make up their vessel and checks the bilges reflexively. Nope, not sinking. It occurs to me that our boat is built for this kind of abuse even if we aren't. Now on her twenty-eighth trip to Greenland, *Bowdoin* is a rugged little ship, built for exploring the Arctic back when much of the place was still an unmapped expanse of rock and ice. Men fished the Grand Banks year-round in vessels much like this one, made to withstand most of the trouble that their crews could find. There is little else to do from here but wait, to sit through the boredom and hope not to encounter any terror in the process. We keep a lookout on deck. We check the bilges. Most of the crew take naps. Late that afternoon someone comes to get me in the chartroom.

"It's moderating," he says.

The following morning all that's left is a sunny day and a lumpy diminishing sea. The barometer rises from the dead and starts sharply upward as the wind dies away. The route to Greenland from Newfoundland is a thousand-mile line, drawn straight north across a piece of ocean that's not well attended to. We've had fleeting information about the weather from static-charged radio broadcasts, a bit more from blurry maps, but we've mostly been figuring it out for ourselves. The gale was not a complete surprise, but our response has been more of a reaction than a plan. I miss the comforting steady crackle of NOAA weather on the radio at home.

It stays nearly calm until our landfall, two days later, when the

Greenlandic skyline appears in the east—a wall built from black mountains, the gaps between filled by descending glaciers and runnels of gray cloud. The arctic sun skips sideways across the sky, and at dawn comes out from behind the peaks in a sudden flash of light like the opening of a furnace door. We all line up to stare, double-wide in our bulky insulated coveralls. The radio comes alive through the open chartroom hatch. A local weather broadcast!

In Danish.

1

THE HOURLY

From where I'm standing, I hear the ship's clock start to ring. It's a sonorous repeating chime, one for each half hour that has passed since the watch began. As the bells fade, the second mate emerges onto the quarterdeck with a clipboard and pencil, excusing herself to brush past me and stand to windward, gazing out at exactly nothing. She looks up at the sky, turns in a partial circle with the unconscious balance of a dancer, and makes a mark or two on a gridded sheet of paper. On her way back to the chartroom, she checks the compass and flips open the cover of a narrow box by the ladder. A birdhouse? One might think, but there's actually a thermometer in there. She vanishes from my view, but a second later there is another percussive note, coarser than the bell, as the end of her pencil taps the rim of the barometer dial.

There is a puff of wind, the ship leans over, and the helmsman makes a half-turn of the wheel to adjust. Nobody says anything. The light goes blurry in a fine spray that seems to be coming from everywhere at once, more like mist from a sprinkler than actual rain. Above us a loose skein of clouds is boiling away to leeward,

a shaft of sun panning the sea surface behind it like a slow spot-
light. The mate is back now. Holding a cup of tea between her
gloved hands, she makes eye contact with an interrogative five-
degree tilt of the head, the pencil stuck behind her ear.

"I don't think we need to make a sail change at this point," I
tell her. "These showers are left over from the squalls that pushed
through before breakfast. We're probably OK even if the wind
builds a bit more. I'm happy to watch things here if you want to
go below and finish up."

It is just another morning off New Zealand, a place where all
four seasons can visit you in the space of an hour. A few minutes
later the helmsman and I are alone again, and I hear the flutter-
ing pages of the logbook as the last bits of hourly data are recorded. Here, north of Auckland, a chain of offshore islands
forms the Hauraki Gulf, a long reach of partially sheltered water
that stretches for about forty miles before petering out into open
ocean—a transition that is marked reliably by the appearance of a
long Pacific swell, which even on fair days gives evidence of distant energies. Today a tropical cyclone near Samoa is sending
smooth hummocks of water toward us from the northeast, two
meters tall. They catch the sunlight and lift our quarter as they
roll past with a subtle greasy motion that rattles rigging and
makes it hard for some to read or work below.

The sawtooth ridges of New Zealand's uppermost peninsula
follow the horizon to our west. Known in the brief Kiwi vernacular simply as "Northland," this is likely the first bit of terrain that
Polynesian navigators saw on reaching the place they would come
to call Aotearoa—the Land of the Long White Cloud. True to
form, crenellations of cumulus are this morning parked above the
hills, their ragged rows of condensation formed as the topography
drives moist air aloft. By afternoon they will be fully developed, a
curtain curving away as far as one can see. In these warm middle

latitudes the deep meanderings of tropical water meet colder currents from the south, rich with lifted nutrients that feed uncountable swarms of marine algae—all packed with oil like miniature avocados or tiny drops of bacon grease. An entire food web emerges from this slippery beginning. Seabirds swarm over schools of bait, while at night the sharks come to graze on squid in the pool of our lights.

By noon it's mild and dry, idyllic. It is one of the many faces this ocean wears in summer, brought by the large high-pressure systems that walk in a slow parade from the west across New Zealand's land mass. It can be delightful under their footprints— think Southern California. One might also think of Southern California in imagining the drawbacks of weather that stays too perfectly sunny for too long. In this, the austral summer of 2020, northern New Zealand is in the midst of a forty-day drought. A thousand miles west across the Tasman Sea, Australia is suffering horrific brush fires, the land baked dry by too many beach days one after the other. Aotearoa has thus far been spared from fire, but is visibly dehydrated even from our seaward viewpoint, the green grass parched to pale yellow.

Rain will come eventually, either from the north in a mass of wet tropical air or from the south as an abrupt and wintry frontal system, an inrushing cold blast from higher latitudes. Below New Zealand a steady west wind howls clear around the globe, breaking against the belly of Australia into waves of atmosphere that roll northeast, near perpendicular to the long massif of the South Island. It is a thrashing eel of subpolar cold. The mountains here, two miles tall, are an anvil to the wind's hammer—forcing the cold air aloft as it advances in walls of rain. The satellite images of all this resemble the sweepings of some giant broom, windrows of airborne debris on a continental scale. Wine and wool towns like Blenheim and Napier sit in sunny shelter east of the ranges while the west gets biblical precipitation. In December of 2019, tourists in Fiordland National Park were stranded as feet of rain closed roads under swollen rivers and collapsed hillsides. It was an extreme event, but in its own way not unusual.

From sea these cold fronts appear as curling buttresses of cloud, like waves in a surfing photo. Before a front arrives, the wind is northerly—therefore warm—and then suddenly it is not. The cloud cliff passes overhead, there is perhaps a lull, and then comes air at great speed from somewhere much colder and more violent. A network of coastal radio stations advises mariners on the real-time progress of these advancing walls of wind—lest they be caught with their topsails set, metaphorically or otherwise. I've spent the better part of three decades going to sea, but a trip to New Zealand, even in its nominal summer, retains for me the sense of a final exam.

Sure enough, a day later gale warnings start to appear in our morning weather forecasts. We go to work securing the ship, the sky now tinted amber with a thin cast of cloud, a halo around the sun. A single petrel keeps pace with us alongside, barely beating its wings. When the wind builds, the air will be full of birds, rising off the water to harvest the sudden abundance of free energy. They own the wind, racing by at eye level as if to be sure we understand. Below the surface are things seeable only when the sea is calm—the dolphins, grazing whales, sharks, and mola, ocean sunfish as big as car hoods. Once a giant leatherback turtle, four feet across with long triangular flippers and drooping dinosaur eyelids. I've seen their babies on a beach in Mexico racing toward the surf, identical but small as silver dollars. These are the things that make a life ashore seem hard to reconcile. They are there all the time, but waves make them vanish.

"Why is this spot called the Bay of Plenty?" someone asks.

I don't remember exactly, other than to say that Captain James Cook named it on his first voyage to New Zealand—or, rather, put title to a place that no doubt had another name already, given by the locals. The Māori had come by water themselves half a millennium before, families packed with their belongings in double-hulled waka from elsewhere in the sprawling Polynesian archipelago. New Zealand is the last place they sailed to in great numbers, the final major land mass on Earth to be settled by humans. Traditional stories place some of their landings not far from our current position, near the present-day city of Tauranga. It's easy to imagine—the pitching canoes at the end of a long and terrifying voyage, captured by the broad thumb and forefinger of Aotearoa. The islanders hauling their gear up the beach—tools, coconuts, a scramble of whatever animals had survived the trip. Ashore they would find trees two hundred feet tall, plus flightless birds the size of alpacas and an eagle large enough to hunt them down. That part is harder to imagine.

THE weather finds us in late afternoon—a dark swath of ripples moving across smooth water, the shearline between two great atmospheric masses. The clouds overhead are long tubular billows, trailing a curtain of rain like steam. Suddenly it is colder. We trim our few remaining sails and keep the wind to starboard, hurrying to close deck hatches as the rain reaches us. Indolent summer is suddenly forgotten.

In an hour we have a gale as promised—45 knots of wind from the south and the sea pushed up into blue-gray hillocks with tumbling foam piles on top. As in my long-ago Arctic odyssey, we lie hove to, one of a few different options available for vessels wishing to ride out weather at sea. Heaving to can mean simply drifting, though for sailing ships a more nuanced technique exists—a balance of canvas and rudder to keep you parked obliquely to the elements, rising and falling like a cork as things swirl around you. With enough room to spare, a ship can sit safely in this condition for some time. Safety does not equal comfort, necessarily, and the ride is still exhausting—comparable to driving a badly sprung vehicle over rough roads for days on end. The wind howls. Doors are impossible to open and then slam shut spontaneously, propelled by wind or abrupt rolls of the ship. Curtains of rain pelt the sea and whistle through the slots between deck structures. Anything that once looked secure seems suddenly fragile, and the process of just thinking about what might go wrong is tiring all on its own.

After dark we permit only a few crew on deck at a time to keep watch, in a noisy black world where the vibrating halyards turn our masthead lights into flickering strobes. This, up to now, is bad weather of the routine sort, the kind from which a ship like ours can expect to emerge without issue if things go well and we are careful. People take turns making rounds of the vessel, scuttling about like crabs to be sure all is secure and to get ahead of whatever problems might be in the offing. A loose line, preparing to slip overboard and foul itself unseen around the propeller—or (equally awful) a gallon of salad oil left unsecured on a pantry shelf, poised to launch from on high and turn a whole room into a deadly skating rink. My friends at home are eager to hear sea stories, but for the most part real tales of heavy weather involve simple endurance—low-grade misery, a constant queasy vigilance in anticipation of some cascading mishap.

As an alternative to just drifting slowly with the flow, ships wishing to maintain steerage can also run before bad weather—putting the wind astern and enjoying the comparable relief that this offers, as the apparent wind is reduced by the speed of the vessel's own motion. This technique has its own drawbacks, most of all something called *broaching*—whereby the vessel skids and slows and is overtaken by a following sea. Helmsmen are sometimes washed overboard this way, or equipment ripped off the deck and carried into the sea as water floods down unsecured openings. Sailing ships once loaded cargos of grain in South Australia and ran for Cape Horn, propelled by the constant storms of the Southern Ocean. There, on occasion, they were swept by waves that tore their deckhouses clean away, sending the surviving officers and any unlucky passengers forward to live out the voyage among the common sailors.

We are on this voyage bound south for the port of Wellington, and then to Christchurch, where a new crew is scheduled to come aboard and sail the ship back to Tahiti. A year ago on a similar passage this ship was caught by a gale that grew rapidly into a hurricane-strength storm, a sudden freezing blast in which the vessel went from comfortably afloat to nearly underwater. This, we would later learn, was probably a *sting jet*—an event typical of a storm variant known to meteorologists as a bent-back cyclone. Such fascinating details are meaningless in the moment. The crew rushed to strike every last bit of sail while the captain—a young man named Jay Amster—took the wheel himself to put the ship before the wind as quickly as possible. Afraid to trust anyone else with the task, he steered alone for six hours until the wind eased back to a mere 40 knots. Several months later I heard him tell his story to a group of meteorologists at a weather conference, in the ultimate meet-your-customer moment. The room was agog, a well-dressed cadre of scientists from around the world getting what might have been their first sailor's-eye view of such a storm, taken off the weather map and transformed into an experience. These are the things that keep us awake in our bunks at night—not the forecast, but anticipation of the unexpected.

* * *

THROUGH each weather event, tempestuous or idyllic, someone will emerge every hour from the chartroom to take down the metrics of all in process. Date, time, wind, waves, pressure, temperature, and cloud cover, all marked on a neat grid in wobbly ballpoint. Compulsive about information, the sailor writes down everything—a ritual that probably reaches back to when some adventurous Egyptian first put to sea with a clay tablet and stylus. Every data point gets meticulously equal treatment, whether it denotes a full gale or calm sunny morning. Mariners have used this method over time to turn the empty ocean into a thoroughfare for travel and commerce: predictable, if never fully known. Magellan set out for the tropics with confidence that he'd find easterly winds there—all because some other intrepid navigator had gone before him and written it down.

A systematic observation of weather lets you have a fundamental sense of what's happening and what might be on the way. Where is the wind from? Is it changing in strength or direction? What are the sea conditions? What about the clouds? Atmospheric pressure is in constant flux, and the rate and trend of change mean everything. The sailor views this aggregation of facts and makes an educated guess at how their day might go. Such has been the nature of voyaging for millennia. Only since the middle of the twentieth century—a period equal to perhaps 2 percent of the age of human seafaring—have we set forth with any notion that someone else has useful advice for us in real time on what might be coming.

Forecasts now happily form the other half of a partnership between mariners and meteorologists—distant creators of the miraculous images that nonetheless remain blurred against the sheer scale of nature. On maps drawn by computers and tweaked by scientists, the location of weather features is projected but never precisely known. Only after careful on-scene observation can a sailor put their ship fully into the picture—rather like street signs might orient you on a city map drawn by someone far away. If you are involved with a cyclone in the northern hemisphere and see that the wind is from the south, you can be sure that the cold front has not yet passed your location—this regardless of what the latest weather map is telling you.

This principle works in reverse for the forecasters, who rely on human measurements for confirmation that their models are telling them the truth. There are nonstop streams of data issuing from satellites, weather buoys, and balloons—but there is nothing better than a live person sending information about pressure and wind velocity at a specific location. "Only you know the weather at your position," we are admonished, in a text sent daily from the National Weather Service. Hence our second mate, and her many deputies at different hours, filling out their boxes with the diligence of clerks in a Dickens novel. The oceans are vast and ships are few, making data points from vessels the most precious ones of all—so much so that weather offices, typically faceless and noninteractive agencies, may take on a pushiness approaching salesmanship if you get on their list of potential data sources.

On our arrival at Wellington an agent of the Meteorological Service of New Zealand strides aboard, semi-invited, to check the accuracy of our barometer and heckle us over the paucity of reports coming from the ship since our arrival in Kiwi waters.

"We've got weethah here too, mate," he says.

I can't argue with him.

The National Oceanic and Atmospheric Association publishes a monthly total of data received from each ship in its voluntary reporting system. It reminds me of middle school, where teachers would post our weekly progress on the board for all to see—a classic motivational tool, applied through a mixture of praise and humiliation. The leaders are celebrated—"Merchant Vessel *Grebe Arrow*: 720 reports!"—and laggards duly shamed—"Research Vessel *Robert C. Seamans*: 21 reports." To be fair, there is an advantage for large ships, many of which have instruments to collect and send the data automatically. Small-craft crew must write it down and transmit it manually during a key interval before each forecast run, a time known as the *synoptic hour*. Today that task involves just a short text message sent by satellite, though for many years the process required a long series of codes read slowly over the radio, as though by a Cold War spy reporting home to Moscow.

∽

THE standardization of weather data into their modern form began early in the nineteenth century, thanks in large part to the efforts of a British naval officer named Francis Beaufort. Prior to then, ships recorded wind and sea conditions with no small measure of subjectivity. To wit:

19 January. We suffered a whole greate gale of wynde on this day, rending the sea into tumultuous spyndryft.

You get the idea. Beaufort was a gifted hydrographer, father to the famed system of British Admiralty Charts that remain the gold standard for navigators to this day. But best known of his legacies is the standard scale for wind velocity that now bears his name, a twelve-tiered matrix binding sea state and wind strength together as a combined measurement. Like most of history's great innovations, Beaufort's eponymous wind scale is as much compilation as true invention. It might instead be called the Smeaton-Dalrymple-Beaufort scale, if equal credit were given to some of its lesser-known progenitors. Whoever deserves credit, the thing is a work of real genius. By correlating wind speed with visual observations of the sea surface, it enables a mariner to make repeatable measurements without any instruments at all—and while later versions include numerical wind speeds to go along with the original duodecimal steps, the Beaufort scale at its heart is a measurement of *force*, not simply velocity.

A captain in his own right, Beaufort aimed to explain what the wind was doing to the water as it blew, thus crafting a sort of universal operating manual for ships at sea. His original version goes beyond mere scalar advice to provide direct counsel on the sails a ship might carry as the breeze builds. At force 6, just halfway up the scale, the wind begins to have a significant effect on the water—and the sailor, from Beaufort, gets some very specific guidance in response: Reef your topsails. In 1831 Beaufort used his newfound influence as hydrographer of the Navy to press his signature scale onto a Captain Robert FitzRoy, master of a survey ship called HMS *Beagle*. While an odd young naturalist named Darwin pored over journals in his cabin, FitzRoy took up the first application of modern marine weather observation. Beaufort spent the next six years convincing other captains to do the same until it was officially adopted by the Admiralty for use on all HM ships. The rest is history.

Wind speed at sea is measured in *knots*, or nautical miles per hour. A nautical mile is 1.15 statute miles, or 6080 feet—a unit equal to one minute of arc measured along Earth's circumference. Beaufort force 6 (a *strong breeze* of around 25 knots) is the point at which wind and sea conditions generally start to become a discernable challenge for small craft. Force 8, a *gale*, is between 34 and 40 knots, and is where marine weather forecasts begin to include direct warnings about wind strength. "Upper topsails shall be struck," Beaufort might advise, or in modern doctrine: "Transits in gale conditions should be undertaken with extreme caution and only by properly equipped seagoing vessels."

The force of wind can be visualized in rough terms using the standard expression for kinetic energy, which (for math lovers) is an exponential square function:

$$F = \frac{1}{2}\, m\, v^2$$

This relationship means that wind force (F) will *quadruple* with a doubling in the velocity (v). A 20-knot wind hence feels four times as powerful as a 10-knot wind. A force 8 gale, only 10 knots above a force 6 strong breeze, will double the pressure on your sails. If you plot this relationship on a graph, it makes a hockey-stick shape, just like the more famous Keeling CO_2 curve in climate science. On the graph of wind force, a gale sits right around the inflection point, where the curve slope transitions from horizontal to vertical. For the average person in town, this would be when simply walking around becomes a real challenge. An urban version of the Beaufort scale might look like this:

- *Force 6: Umbrellas ruined.*
- *Force 8: Café furniture blows into street; pedestrians cannot walk in straight line.*
- *Force 10: Don't go out.*
- *Force 12: Call 9-1-1.*

～～

WIND transmits energy to the sea surface through aerodynamic drag, making waves that grow larger through a set of linked factors. Ultimately, stronger winds have the potential to make the biggest waves. Also important are the wind *duration* (how long winds blow) and *fetch* (the uninterrupted distance over which wind travels). Harbors invert the principle of fetch to provide shelter: There may be a lot of wind, but over short distances it can't do much with the water. On the other hand, wind blowing unimpeded over long spaces can build waves very quickly. A 40-knot wind blowing across ten miles of water might make waves with a height of seven feet, while if the fetch is unlimited (the effective condition for oceans), the heights can quadruple, to twenty-eight feet.

Waves are measured according to their height, from peak to trough, and *period*, which is the time in seconds between successive wave peaks. Multiplying the period by three gives you the wave speed in knots, a trick that forecasters can use to determine whether seas will continue to build in some cases. Waves only grow while they are involved with the wind that's forcing them. Once the wind stops, waves persist in the form of *swells*, retaining momentum for what can be thousands of miles. The most energetic waves travel farthest and last the longest. Surfers listen assiduously to data from offshore buoys, watching for signs of the long-period swells that will come ashore as monster breaks.

The motion of storms themselves adds another layer of influence. Fast-moving weather systems are often connected to rapid shifts in wind direction, meaning that their waves may be smaller but much more chaotic—and therefore more dangerous—as a cross-sea develops. Forecasters also watch carefully for cases of so-called *dynamic fetch*, created when a storm keeps pace with the waves it is making—effectively staying with them until they have achieved their maximum potential size. The detailed connectivity between wind velocity, fetch, and duration is spelled out in a fascinating graphic called the Sverdrup-Munk-Bretschneider wave nomogram, derived by three giants of oceanography in the mid-twentieth century. There are a lot of lines in this drawing, and it's not something that you whip out to consult on an hourly basis, but the relationships it lays out are easy to generalize: As the wind blows harder, longer, and farther, the waves grow taller and pick up more speed. In the end the growth of waves is limited by their velocity. Once a wave is moving as fast as the wind, it can get no larger.

It's most common in offshore environments to experience several wave systems at once, as seas from the local winds mix with swells left from past or distant events. It can often be hard to tell just what you are seeing. Waves may add together through a process called *constructive interference* or cancel one another out by *destructive interference*. After sudden wind shifts the pattern of seas can quickly become chaotic. Regardless, waves are never perfectly uniform, and sea heights are generally given as the mean value of the tallest third of waves. Some weather agencies favor categorical terms over numerical ones, and I've become fond of the somewhat droll British descriptors: *calm, moderate, rough, very rough, high, very high,* and . . . *phenomenal*. Recognizing the drawbacks of the phenomenal, forecasters try hard to give notice of cases where extreme conditions are likely to develop.

A boat sitting on a deepwater wave will bob up and down—traveling on a path that is actually circular in profile. When waves reach shallow water, things happen much differently. Here friction with the bottom slows the lower part of the wave, causing the face to steepen and eventually collapse. This is a critical transition, the point where a keg of gunpowder becomes an explosion. When waves break, the water itself sets into motion, traveling at a speed proportional to the original wave train. This dissipation of energy can happen slowly—as a wave runs up onto a gently sloping beach—or all at once—as a long ocean swell strikes suddenly on a reef face. You can stop a speeding car by letting it roll slowly up a hill or by driving it into a brick wall. Navigating through breaking surf is one of the most dangerous and specialized shiphandling tasks of all, and certain waterways—most famously the Columbia River entrance between Washington and Oregon—are at times closed outright by wave action in their approaches. Deep-water waves can also break in some cases, particularly during interference with other wave trains. Here the hazard to small craft can be similar to that of a surfer caught inside, forced to absorb the impact from tons of water cascading at great speed.

I'm often asked what the biggest wave is that I've ever seen. Outside of a boat ride I once took to watch a surfing event in Tahiti, the best answer I can give is, I'm not sure. Anybody who goes to sea for a living has probably seen twenty-foot seas on a regular basis, and thirty-footers on occasion. There are a few historical contenders in play for the title of tallest wave ever measured, but it is generally accepted that waves of sixty feet are not uncommon in certain oceans. There's also some self-selection at work. Sailing vessels and other small craft plan passages to avoid heavy weather where possible. Large commercial ships have the power and strength to endure all but the worst conditions, and as a consequence will often sail regardless—a practice that works well until a flawed judgment is applied at the wrong time.

The TV reality show *Deadliest Catch*, with its live footage of Alaskan crab fishermen at work in the Bering Sea, has allowed people in the safety of their living rooms an unprecedented view of ships working in absolutely horrible conditions. The most frightening story of waves I've ever heard might actually have come from my friend Rick Fehst, who's been captain of several boats in the Alaska fishery. Hurricane-strength winter storms are common in the Bering Sea, as frigid, bone-dry air blows across Siberia and piles into the warm currents of the western Pacific. For Captain Rick, things are typically the worst on the back side of these systems, where the wind shifts to the northwest and begins to roar in earnest. The snow may be finished by this point, but the change in wind direction creates a treacherous cross-sea in which no two waves are the same.

Different ships have different strategies for survival in such storms. For a crab boat heavy with fuel and gear, I learn, the preferred technique is to steam directly into the wind and waves, maneuvering to meet each sea bow-on. In 70 knots of wind, this is not a job for amateurs, and Rick explains what's necessary to get it done while he tends to some salmon fillets on his porch barbecue. He is about my age, with the slightly ursine aspect of many career fishermen that I know—a kind of well-insulated sure-footedness. You can, I suppose, be slender and still spend a life chasing fish in cold water, but some thermodynamic scale is tipped immediately against you.

"In thirty-five- or forty-foot seas," says Rick, "you need to square up on each one and go full speed ahead to make sure you make it up the face of the wave. Sometimes the tops are breaking, and then you need to go full astern so you don't hit the foam pile too hard and get your wheelhouse windows blown in. After that you get down on your knees and duck for shelter behind the bridge console. There have been skippers blown right out the back of their wheelhouses and over the side in stuff like this. Then you're back up on your feet and giving it full ahead to get lined up for the next wave, which may be coming from a different direction. I've spent up to thirty-six hours at a stretch in the chair doing this and having the crew bring me coffee and sandwiches to keep me going."

It is a riveting account, equal parts survival and my-day-at-the-office. Rick drifted up to Alaska in his twenties, looking for money to pay for film school. In a pattern that I recognize from my own career, some latent aptitude and the year-on-year accrual of time turned a detour into a vocation, obscuring whatever career plans had gone before. Now boat owners compete for his services, and his wife, Ann Nora, is managing physician of the medical clinic in Dutch Harbor—where her specialty is repairing damaged fishermen so that they can return quickly to their paid occupations. After lunch we go inside to look at drawings of their new boat, a tiny steel tank of a yacht they plan to use for charters, taking guests on for views of an Alaska that few get to see. As far as I know, Rick is still waiting to make a movie.

⌐~⌐

On her first trip into space, NASA astronaut Sally Ride looked out and marveled at the frailty of the atmosphere. It reminded her of the fuzz on a tennis ball. Three hundred miles—a generous measurement of the atmosphere's overall thickness—is just 0.025 percent of Earth's diameter. The skin on a mango. Even if you add in the ocean, the liquid base of the air-water laminate that makes life on Earth possible, you're not left with much more. And yet somehow we live here, unlikely colonists in a turbid membrane that lets us breathe and holds our future.

The air in your lungs is four-fifths nitrogen. The rest is mostly oxygen, with a whiff of carbon dioxide, methane, and argon thrown in. There's also water vapor—the invisible gaseous phase of H_2O—evaporated from the ocean and floating around in ratios as high as 4 percent. When air cools below a certain temperature, water vapor condenses into clouds—airborne scatterings of liquid and ice that are the only visible part of the atmosphere. As well as giving us oxygen to breathe, the atmosphere stabilizes the temperature on Earth by retaining solar energy. Consider our nearby moon—with no atmosphere at all—where it is 253 degrees Fahrenheit by day, and minus 387 by night.

The atmosphere is held close to Earth's surface by gravity, most of it compressed into a dense lower layer called the *troposphere*. The troposphere has 80 percent of Earth's air, along with nearly all of its water vapor. An upper boundary, the *tropopause*, is the limit of most terrestrial weather processes. Jets like to fly just above here in the *stratosphere*, where thin stable air makes travel fast and smooth. Weather is occasionally an issue up in the stratosphere, but only rarely. The relative height of the tropopause depends on temperature. Because cold air is so much denser than warm air, the troposphere is a scant four miles thick at the poles and twelve miles thick near the equator. Mountains are as a rule easier to climb in the tropics, since you're able to go higher before you run out of atmosphere.

I am discussing this point with a scientist on the ship.

"Ah, of course," she says. "The peanut M&M theorem."

I am at a loss. I listen carefully.

"The chocolate," she continues, "is always thicker in some places than in others."

"Well . . . exactly."

BAROMETERS work by weighing the atmosphere as it sits on top of us. At sea level, this weight yields an average pressure of 1013 millibars, or just over fourteen pounds per square inch. Pressure *decreases* at higher altitudes, since less air remains overhead to push down on your barometer. Because air is compressible, this decline happens at an exponential rate, very quickly at first. A climber high on some Himalayan peak might measure a pressure of only 500 millibars, half of the standard surface value. It's hard to breathe in such places because the force of air pressing into your lungs has been reduced by 50 percent. A jet cruising at 35,000 feet—in the lowest reaches of the stratosphere—has a scant 200 millibars outside its cockpit, with just a fifth of the atmosphere's mass remaining above.

The Galilean protégé Evangelista Torricelli developed the first known barometer in a series of experiments done around 1643. Torricelli constructed a long glass vacuum tube, closed at one end and partially filled with a slug of mercury—just the sort of thing you could do before industrial safety standards came to science labs. With the open end set in a dish that he'd also filled with mercury, Torricelli saw that the column of quicksilver in his tube would change height slightly from day to day—a trend he correctly attributed to variations in atmospheric density. Until the invention of the dial-gauge barometer in 1844, all the instruments that followed were some variant on Torricelli's theme: a liquid, commonly mercury, captive in some transparent vessel that was open at one end to admit the force of atmospheric pressure. Mariners paused each hour to read this delicate tool, a precious device known colloquially as "the glass."

Hourly readings of pressure don't mean much by themselves; it is the trends that are significant. A sharp rise or fall in the barometer indicates a trend of changing pressure—a so-called pressure *gradient*—and with that comes wind. Science teachers demonstrate this principle by letting the air out of a balloon, showing how a gas will flow from an environment of higher pressure toward lower pressure until the differences are equalized. Lower atmospheric pressures are typically connected to rising air and the subsequent condensation of water vapor into cloud droplets. For this reason a falling barometer is often linked with the advent of clouds and precipitation. Conversely, high pressure is usually a product of stable descending air.

A high barometer may portend calm at your location, but not always. A forecaster in New Zealand once told me that any zone of high pressure in excess of 1030 millibars could be counted on to have gale-strength winds somewhere on its edges—rather like a very tall mountain will be surrounded by cliffs. "Over 1030, somewhere it's dirty!" he said, not without some satisfaction. Low-pressure systems tend to have their steepest gradients near the center, so if your barometer reads much below 1000 millibars, it's probably already windy where you are. At any rate, rapidly changing pressure is always a red flag for wind. Meteorologists enthusiastically categorize a one-day drop of more than 24 millibars as a *bomb*—a virtual guarantee of storm conditions for anyone involved.

⌒

THE last space in the weather log is reserved for the observer's initials, just after a paired set of columns for recording the type and quantity of clouds in the sky. There is sadly no easy-to-read instrument for naming clouds. Typically, one begins by using a pie-slice system to divide the sky into sections, yielding a rough estimate of what fraction is covered. After that comes the identification of specific types. Here the sailor, like a bird-watcher, is left to make judgments based on what they see, often while consulting a printed guide or receiving advice from their compatriots, sometimes helpful:

"Is that a bobtailed scarlet bunting?"

"Absolutely not! They have pointed wingtips and rarely fly that high."

Like Beaufort's wind scale, the modern taxonomy of clouds dates to the early nineteenth century, a period in which the gentlemen of the West were on a seeming bender to measure, identify, and name everything they could find in the natural world. In the case of clouds the chief agent of this process was Luke Howard, a British pharmacist who wrote extensively on meteorology. In his 1803 *Essay on the Modification of Clouds*, he penned the following:

> *Clouds are subject to certain distinct modifications, produced by the general causes which affect all the variations of the atmosphere; they are commonly as good visible indicators of the*

operation of these causes, as is the countenance of the state of a person's mind or body.

Howard pioneered the use of a Latin-based Linnaean system of naming for clouds, as had been established elsewhere in the natural sciences. We have Howard to thank for the three main categorical groupings—*cumulus, stratus,* and *cirrus*—as well as the system of prefixes and suffixes that permits the sorting of objects that are by their nature in constant flux. Just as with Beaufort's tabulation of wind and sea conditions, sailors were suddenly able to take a flying leap from the subjective to the systematic. A "skyeful of greate sheep and smoothe fields of cloude" thus became "⅝ cumulus with some stratus," information that anyone with the same training could visualize and apply. In likening clouds to the human countenance of the atmosphere, Howard was onto something. Rather like facial expressions, clouds can tell a story quite plainly or leave much in doubt.

I T is thirty-six hours before the wind diminishes enough for us to resume our progress toward the south. The tall clouds of the cold front are followed by a day of grim overcast, bands of showers, and a wind that feels the utter opposite of summer. There is nothing, I am reminded, between New Zealand and Antarctica. The barometer tells the story of what's come to pass, a slow and tepid descent followed by a brisk rise as the cold dense blast of subpolar air makes its arrival. During this time we have drifted backwards fifty miles, lost ground that is shown discouragingly on the damp chart by a wandering regression of pencil marks. Somebody has drawn a frowny face on a Post-it note and stuck it next to our plot. This sort of levity is discouraged in navigation, but it's fair to say that everyone feels the same way about the last day and a half.

There is an email on my computer from our home office, where through the miracle of microwaves our windy backslide has been observed from eight thousand miles away. Is everything OK? Are people feeling well, and is the ship still on schedule? The planners want to know. Over my head is the controlled clamor of the mainsail being set, subtle accelerations as canvas fills and the ship gathers way. The ancient sensations of sailing are apparent even as my backlit screen shows that—unlike the original mariners of this ocean—we are alone here but no longer truly by ourselves. In imagining the Polynesians and their adventurous leaps across the Pacific, I consider the more recent voyages that have brought our own ship back and forth across these same waters—a modern steel vessel a hundred times more robust, connected by satellite and surrounded by a sense of the known. My scientist friends tell me of all the ways that the ocean and climate have changed since people first put to sea, but for a sailor on an open deck these trans-formations remain largely an abstraction. The flashing cursor and keyboard aside, it is hard here not to feel some connection with all the others who have crossed this ocean previously. Pacific voy-agers, captains of discovery, the traders and mapmakers—surely all marveled at the same endless show of light and clouds. No doubt many were at times as cold and wet as us, and just as baffled by the unexpected.

2

FIRST PRINCIPLES

B ien viaje," says my driver.

He pockets his fare and reaches to turn up the radio, glancing quickly back at his mirror for traffic.

The ship is waiting next to us at the wharf, a square-rigged anachronism found easily amid the waterfront forest of palm trunks and aluminum yacht masts. I nod to people with half-familiar faces and bring my bags up the gangway—a routine transit made since my twenties, still with the same small sense of incipient vertigo. A voyage begins.

The *Robert C. Seamans* is forty-two meters long, a sailing school ship built of steel and certified to carry a crew of thirty-eight on any of the world's oceans. She has white topsides, tan spars, her gear well-kept but with the characteristic patina of working vessels. Her name is displayed on trailboards at the bow, raised wooden plaques that have from time to time been lost to the sea in severe weather. Robert Seamans himself was a star in the golden age of aerospace engineering—a NASA director during the Apollo program and later secretary of the Air Force.

When his family granted funds to build a ship in his honor, devoted to scientific research and education, he reportedly said, "Fine—just don't name it after me." One can prevail in arguments with presidents, but others are unwinnable.

The *Seamans* and a sister vessel work together under the banner of the SEA Education Association, a small research organization based in the scientific mecca of Woods Hole, Massachusetts. I have a cluttered cubby of an office there, down the hall from a lobby full of maps and ship models and some classrooms where we prepare students for their first experience at sea—as though such a thing were truly possible. The association has kept me busy for large sections of my sailing life—a compelling occupation, if not a lucrative one. Over the years I have with my overlarge duffel flown untold miles to meet one ship or the other, stepping aboard to join the ever-repeating play of departure.

"You are leaving for a trip soon?" my mother asks me. "Atlantic or Pacific?"

"Let me look at my ticket."

~

I T's the best time of day in Puerto Vallarta—just before sunrise, when the mountains around the city are backlit with orange and the rest of the sky is deep blue. By the terminal gate a video billboard still twinkles, set behind silhouetted vegetation like some low-latitude fragment of Times Square. The harbor basin has doubled in size since my last visit, blasted out of the mangroves as the city chases its Faustian bargain with the cruise ship industry. Now there are three full-sized berths, rimmed with neat plots of cut grass and smooth asphalt. Freshly built retail properties are situated nearby to extract money from roving passengers: The Galerías Vallarta Mall. The Hard Rock Cafe. Denny's. The port is hopeful that this bloom of construction will lure cruise lines into using Puerto Vallarta for a base, increasing traffic and revenue by some yet unknown multiplier. One senses the unfortunate gestation of a new Waikiki, built here on the shore only inches from the Mexico that came before it. In town two dusty vaqueros pick their way on horseback through the madness and tie their mounts outside the grocery store, on some routine errand first run long before this all began.

The captain I've come to relieve is waiting aboard, bags packed and surfboards already strapped to the roof of her rental car. We'll be here together for a day, but the weighty handoff of an $8 million vessel and its crew—what another captain I know calls "the sphere"—is surprisingly brief. A run-through of papers, discussion of a minor issue with the number two generator. A report on the stores being delivered today: two months' worth of food for forty people, all of whom will be eating voraciously once their seasickness abates. We solemnly unlock the safe and count the cash. There are 5000 US dollars, some Mexican pesos, and a medley of colorful island currencies in a worn envelope, exotic promises of places we may go. A box of morphine ampoules that I pray to never open. On ships, as in corporations, it is the vice presidents who do the substantive work—the first mate, the chief engineer, the steward. Captains count money, fill out forms, and bear the onus of awaiting the unexpected.

Today there are blessedly few intrusions from the world of the unexpected. Our produce delivery arrives in a truck, 3000 dollars' worth of apples, mangos, long beans, and round green squash rolling on the grass like goblins' heads. All of this will have to last as far into our trip as possible, and to thwart decay every pepper, pear, and bunch of carrots is unpacked and spread out to dry before getting its own personal wrapper of newsprint. The interim rainbow of arrayed fruits and vegetables is beautiful to behold, always intriguing to passersby. The steward checks her clipboard and chats with a dockworker who has stopped to find out exactly what is happening.

"Hola, chica! Qué hacen?"

"El capitán quiere una ensalada muy grande hoy."

"Ah, bien! Siga."

Someone by the mainmast is grinding loudly away at steel with a power tool, sparks flying out in a shower toward the ersatz salad bar on the wharf. I have no idea which among the thousand tasks on our list they are performing. Every ship departure is a jigsaw of critical pieces, all to be completed or abandoned by that most definitive of deadlines, sailing day. Fuel. Water. New starting batteries for the main engine, sixty fresh pillowcases, a hundred dozen eggs. The oceanographers need blue grease pencils, which don't seem to exist in Puerto Vallarta.

Many of these details fall to our agent, who stops by as the eggplants are being loaded aboard. Agents are fixers of the maritime world, accomplices in connecting the polyglot society of ships with the local machinery of the ports that they visit: customs officials, harbormasters, stevedores, technicians, and vendors. Some harbors will not let you work without an agent; others merely make you sorry if you don't. It is an old and clannish profession, tightly held. At the Portuguese island of Madeira our man David wears Italian suits and brings cigars and brandy to the ship as premiums. Here, on the edge of the Mexican desert, agent Juan appears on a tall Honda dirt bike with his call sign—"Paperman"— painted on the tail case. We say hello and share the day's list of business. I hand him our passports for the immigration office, the precious stack counted twice before he roars off in a blue cloud of exhaust.

A day later we leave the dock at first light and anchor off the beach to hold drills, another time-bound obligation in the routine of departing ships. By afternoon there will be a sea breeze forced by warm air rising off the land, but for now it is calm as glass. Two early rowers in a shell stroke their way across the channel like a water strider, a thin line of wake and puddles vanishing behind them. A school of young eagle rays is browsing at the surface nearby, strange waffle-sized fish with long whips for tails.

A grim retinue of possible disasters awaits a vessel at sea, all to be faced in the unaided isolation of the ocean environment: Fire. Flooding. A crew member lost overboard. Sailors are their own first responders, their own lifeguards, and if all else fails must direct their own evacuation from a stricken vessel. I explain this to the trainees, seeking to express the seriousness of it all without provoking panic in what someone once called my kidding-but-not-really voice. At the heart of organizing for these contingencies is something called the station bill, a document that divides the array of threats into a comforting matrix and assigns everyone a job, from the chief mate to the green cadet. Work inhibits panic. We sound alarms and walk through our scripted responses. I check radios with the third mate while two students drag a long snake of fire hose to the rail and wait for water to arrive at the nozzle. On the waist-deck a team is trained to hoist and deploy the rescue boat. Finally, the watches muster by the life rafts, which are rigged in special canisters to launch automatically if time runs out to deploy them by hand. Everyone dons bulky immersion suits and responds in turn as the roll is called. We're close enough for hotel guests to paddle out in kayaks to watch us. Some ask if we need help.

The chief engineer comes to share his concerns over the condition of our second fire pump. A new gasket may solve the problem, but he will let me know. He's also curious about his station bill assignment—should he and the first mate work together on the firefighting team, or separately in case something goes badly wrong? This might provide flexibility in the event of something truly serious, like a fuel fire in the engine room. Dressed for work in a filthy oxford shirt, David has sharp semi-handsome Anglo features and a perched flop of straight dark hair—a reincarnated Edmund Hillary, or perhaps the actor Hugh Grant after a life under harder circumstances. It is hot now, and somewhere in the background I can hear mariachi music playing from the beach.

Engineers are the most important people at sea, even on sailing ships. I would cast my second vote for cooks, for in truth you could live longer on peanut butter sandwiches than you might with a failed power plant or ruptured sewage main. The requirement to rise at any hour and fix whatever is broken calls on a unique family of polymaths—fascinating, if not predictable. David lectures on Russian history while dismantling outboard motors, interspersing stories of far worse days that other people have had on their own jobs. In a previous billet he once watched a tugboat capsize and sink while it was turning to come alongside a wharf. The crew had forgotten to close an important valve and the resultant surge of fuel between tanks was their doom. David's nickname, Danger Dave, derives from another story involving an old girlfriend that I have not yet heard.

The galley has questions also. There is cake with the morning coffee, likely to vanish if we don't hurry. Are we almost finished, or would we like them to save a piece for us? Duly alerted, I walk back to the salon, images of pastry and engine room fires shuffled in my head, impossible to reconcile.

By afternoon we are sailing, the blank spaces in the rig filled by taut white parabolas of canvas and the ship heeling gently in a moderate breeze. The long Pacific swell begins to affect our motion, soon alerting us to any deficiencies in our stowage plan. Some unfindable object rattles randomly in the compartment next to mine. The steward comes to verify that the pantry is fully secured for sea—boxes of pasta and bags of flour swaddled in plastic and tied down tight, a thousand cans wedged together like paving stones. This is serious business. In 2002 off Long Island the schooner *Ernestina* began to flood through a leaky hull seam and nearly sank, an emergency made even more dire when her pumps clogged. The culprit? Chocolate chips, spilled from their container and loose in the bilge. The captain turned her ship decisively toward shallow water, aware that an intentional grounding might be necessary to prevent a total loss—one of the rare cases in which letting your vessel touch bottom is considered good seamanship.

There is plenty of food, to be sure. Stacked in the freezers and every available locker are two hundred pounds of chicken, twenty cases of canned tomatoes, and forty jars of jam. Also five hundred corn tortillas, fifty kilos of butter, and an undisclosed quantity of ice cream. Eating this will be thirteen crew and two dozen college students, the latter for the moment mostly ill and struggling to make sense of anything they are hearing:

"Set the mainsail! Hands to the topping lift, stand by your sheet and halyard!"

"Cast off the downhaul and keep your feet clear of the bight!"

"Wake the idle watch for lunch, and don't forget to plot our noon position!"

Stunned and eager, they rush to help, faces bearing the telltale signs of sensory overload and the glaze of freshly applied sunscreen. Even in these early moments the students are being asked by their watch officers to steer, navigate, keep lookout, and check machinery—all tasks critical to the safety of the ship. Such is the nature of training vessels.

Land recedes behind us at the stately pace of 5 knots. A ten-story structure is visible from about twelve miles away on a clear day, and by midafternoon the hotels along the beach are just dots shimmering on the horizon. An hour later they have vanished, the last terrestrial objects we will see for a month. We are sailing southwest, outbound from the Mexican mainland just below the terminal protuberance of Baja California. Our next landfall will be the island of Nuku Hiva, in the Marquesas group—a far-flung bit of French Polynesia, almost three thousand miles away. In between, only water. There is a light wind from the northwest, dry and almost cool. Below deck it is quiet, with just the hum of ventilation fans and the occasional *whoosh* of a toilet being flushed. The *Robert C. Seamans* could be a very large, slow airliner on its way to a tropical destination.

The north wind dies over the next thirty-six hours, the water becoming smooth as mercury. Shearwaters carve turns across our wake, wingtips barely clear of the surface like a surfer's extended hand. The birds that most people equate with the ocean—the gulls, terns, pelicans, and plovers—are in fact coastal animals. The pelagic world is inhabited by a different population, species that spend their lives at sea and which you are unlikely to encounter unless you go offshore yourself or brave the foul din of some remote nesting colony. These are the shearwaters, petrels, boobies, and stately albatross—the latter gull-like in appearance until the point when they fly close and you realize their wings are eight feet across.

By our third day at sea it is humid, the morning quickly warm as tall piles of cloud bloom in the sky. There is wind now from the east, at first fitful but by dinnertime 15 knots—Beaufort force 4. The sea surface a mosaic of moderate waves with patches of foam showing randomly at their crests. The ship gathers speed, an easy rhythm to her motion—5, 6, 7 knots. We could set more sail and go faster, but at this early date we don't. Few things are more unsettling than too much sail on a dark night, all while the trainees are still learning to work the pencil sharpener. This wind will be with us tomorrow—I am confident of that.

⌒

HERE at the threshold of the tropics we have joined the intake draft of a vast planetary engine, driven by the surfeit of heat near Earth's equator. If you've been to the tropics, you have doubtless noticed these winds, blowing warmly and with little respite from the east. Palm trees lean together in the same direction, and settlements favor the western flank of mountains, where it's quieter and a little less rainy. These are the trade winds, which have, over centuries, swept whole populations of sailors across the wide latitudes—Columbus's first landfall in America was the Bahamas, not because he was looking for them in particular, but because they were there, directly downwind of his fleet in a location that sailors would describe as *to leeward*. Until steam replaced sails, any long voyage west was commonly begun with a transit to the tropics, a belt of fair breezes to carry a vessel easily toward its destination.

Earth warms its atmosphere from below like a skillet cooking a pancake, reemitting heat that the planet has absorbed from sunlight. Wind is a product of the basic inequalities in this process, entrenched deficits that the forces of nature are working ceaselessly to make whole. The tropics, subject to near-vertical insolation, get warm and stay warm. In fact, the regions between the 37th parallels of latitude—say, roughly, from San Francisco to Auckland—operate at a heat surplus, absorbing more energy than they can radiate. Higher latitudes run in the minus column, as Earth's axial tilt alters the sun's place in the sky over the year, leading to great variability between seasons. New Yorkers thus glumly write checks for both heat and air-conditioning, giving up precious closet space to bulky garments that must then hang abandoned from March to October. The poles stay mostly cold, as the low angles of arriving light reduce the density of rays reaching the surface. That's when the sun is up at all, which for large parts of each polar winter it is not.

Warm air rattles its molecules around more briskly, and consequently takes up more space. Air that's been heated becomes less dense and tends to rise, encouraging the arrival of cooler and denser air as a replacement. This buoyant lifting of warm air is known as *convection*, and the inflow of cooler denser air into the resulting void is called *convergence*. Witness both from the comfort of your home with a simple experiment: Open the front door on a winter day and feel cold rush in as precious warmth rises away toward the ceiling. With a sensitive barometer you could perhaps also measure a drop in pressure as the warm air ascends around you. Then open some windows and feel air blow in from multiple directions, for a closer approximation of how this might happen in the atmosphere—a place less constrained by doors and windows.

Warm air can carry more water vapor, since the process of evaporation depends directly on a supply of heat. As water warms, its molecules accelerate, and a greater fraction carom off into a gaseous state. H_2O is the lightest of the atmospheric gases, meaning that humid air—perhaps counterintuitively—is less dense than dry air at the same temperature. The movement of water through its three phases is known in science as the *water cycle*: Liquid H_2O evaporates from the ocean and is carried invisibly through the air, until it undergoes condensation and returns to the surface as rain or snow. Snow melts and runs into the sea or turns directly to vapor through a process called *sublimation*.

Heat energy is transported around the world through these combined actions. Warm humid air becomes buoyant and rises, expanding like a balloon and releasing heat as its cargo of water condenses from vapor back into liquid. Cool air grows denser and sinks, converging toward the spaces left by rising warm air. Air moves horizontally across gradients in pressure, from high to low—a phenomenon known scientifically as *advection*, but recognizable to most people as wind. It's a cyclical process, independent of scale—one that will persist as long as a source of energy remains. A puffy summertime cloud is usually the result of convection. So is a hurricane.

Deep in the tropics, the warmest, wettest air on Earth rises up and is replaced by wind converging constantly from opposite hemispheres. Air masses collide like continental plates to build a mountain range in the atmosphere. This is the *intertropical convergence zone*, or ITCZ, a feature you might think of as the thermal equator of the planet. It's a region of perennially low atmospheric pressure—visible from space as a ragged and lofty encirclement of clouds, tall enough that airplanes must sometimes fly through, not over them. This great belt waxes and wanes, wandering on and off continents and drifting with the seasons. The ITCZ is where it rains the most, and you can find its footprint on maps by looking for the deep green swaths of equatorial rain forest. Sailors at sea historically have called this busy and unpredictable region the "doldrums," but that moniker is deceptive. As we'll see, there is nothing really dull about the doldrums at all.

Surface air flowing toward the ITCZ makes the trade winds, and their strength from season to season depends on the extent of temperature differences between latitudes. The trades may blow gently, or they may howl—particularly in winter, when polar air is coldest and the resulting pressure gradients are steep. Caribbean islanders talk of the "Christmas winds," a time of predictably spirited sailing that may run well into January. This is also the heart of holiday yacht charter season, and radio traffic can be lively as vacationers in their rented boats are heard, dealing with too much of a good thing.

Tropical island chains are often strung together from the

abrupt summits of undersea volcanoes, and the trade winds hit them like a fire hose on a picket fence—funneled at times to twice their normal strength in the gaps. The Canary Islands are famous for this, as is Hawai'i. The Alenuihaha Channel, between Maui and the Big Island, routinely sees winds of 50 knots. You can find yourself fighting for your life on an otherwise perfect sunny afternoon. This is what happened to Captain Cook in February of 1779, when he sailed away after enjoying a month of abundant hospitality at the hands of local residents. It was not a calm day, and when his ships started to self-destruct, Cook retreated, seeking shelter at Kealakekua Bay, near modern-day Kona. It didn't go well. Like so many who have seen off long-staying guests, the Hawaiians were not happy to have the captain and his needy horde return. Things went downhill quickly, and within a day Cook and four of his marines were dead, killed in a skirmish over a stolen longboat.

Sailors in the run-up to Cook's time understood that the trade winds were driven by convection but were less sure about why they came from the east. Why not from the north, simply straight from the poles toward the equator? The English lawyer-scientist George Hadley is credited with giving the first correct explanation, an early version of what ultimately became known as the *Coriolis effect*. Hadley posited that Earth's rotation gave an apparent twist to airborne movements when measured from the ground—so that as wind converged on the tropics, it was deflected into an easterly flow, at least as far as ships could tell. Hadley was right. The Coriolis effect applies to anything moving freely above Earth's surface and owes to the fact that the ground itself is in a rotating frame of reference. Thanks to the principle of angular momentum, a spot near the equator effectively travels farther and faster in each instant than one that's farther north or south. This momentum is conserved, so that things traveling in the atmosphere will take what appears to be a curved path for an observer on the surface—bending to the right in the northern hemisphere and to the left in the south. There is a popular demonstration video of this phenomenon, an overhead view of two children throwing a ball back and forth on a spinning merry-go-round. For them, the ball makes each trip in a straight line, while from above its path is wildly curved—on a trajectory that any baseball pitcher would sell their soul to achieve. The Coriolis effect won't bother your tennis serve but applies in rocketry ballistics and is a major factor in global wind circulation. Regardless of what you've heard, bathroom fixtures are not affected. I've been to New Zealand and seen toilet water swirling in the very same direction as at home.

The trade winds will be our motor for the next month, a steady breath from just over your right shoulder as you stand at the wheel to steer. There's a weightless feeling to sailing this way, evocative of gliders or hot-air balloons. Waves roll up from astern and travel smoothly past. We are having a much better time of it so far than Captain Cook did. A 20-knot breeze blowing across open ocean will eventually build seas ten feet high—not trivial, but the waves for now have a regularity born of winds coming day upon day from exactly the same direction. Think sand dunes instead of ski moguls. Flying fish skitter off the crests, tails grazing the water before rebounding aloft for improbable distances. Attracted to light, they fly aboard after dark and thrash about the deck like great beached cicadas. You can order them for lunch in Barbados; they don't taste bad at all.

﹏

THE day on ships is broken into watches, like shifts at a factory but uncoupled from any other part of the rational calendar. Three teams alternate, ad infinitum, to do all the required work:

navigation, sail handling, running the laboratory, putting out meals, and tending to machinery. Long intervals of routine are broken by bursts of activity, some scheduled, others reactive. In each watch team on the *Robert C. Seamans* are eight trainees, one mate, and one scientist—subparts of a larger colonial organism with a changing set of faces. Six hours on, twelve hours off. Minutes crawl by; days disappear. Over my bunk is the quarterdeck, home to the helm, compass, and most of the watch's routine business. Through the hatch I hear random snatches of conversation, some quite informative:

"Is today Sunday?"

"I have no idea. What's for lunch?"

THE oceanographers collect samples with instruments sent overboard on rope pennants and long winch wires. Our ship must heave to and drift for these procedures, and the mate is discussing this with his team—a bunch of nineteen-year-olds who've suddenly found themselves in the position of sailing a tall ship to Polynesia. *Robert C. Seamans* is rigged as a brigantine, an old design with large triangular fore-and-aft sails for working to windward and a stack of squaresails for running with the breeze astern, as we are now. The watch sets to the baroque and arduous task of bringing the ship about. There are braces to haul, lines run aloft to rotate the long horizontal spars known as yards. Inhauls and clewlines to take in our topsail, now a hindrance as the ship turns into the wind. Sheets, tacks, brails, and preventers. Helm commands. A tangled choreography of steps proceeds, a troupe of neophytes acting under the calm cadence of expert instruction. Five minutes later the ship is stopped, riding easily over the waves like a duck. An array of neatly coiled lines becomes a pasta factory.

Chief mate Jeremy Law supervises the restoration of order, a model of tranquil competence. While in the Coast Guard he was first officer aboard the cutter *Eagle,* one of five sailing ships launched as training vessels by Nazi Germany and ceded to Allied forces after the war. A masterpiece of ship construction, *Eagle* was built in a part of the world where the craft of sail reached its zenith, and its true mastery still resides. At ship gatherings and conferences, I meet the descendants of this heritage: the same gruff Germans, knowing Danes, and tall Dutch who manage to look something like pirates and fashion models all in one. Aloof in their expertise, they go outside in groups to smoke, returning to spring with wry antagonism on other, unsuspecting sailors:

"Why have you rigged it like that? You know that way is wrong, of course."

"Well, yes, it is no problem to do this if you know what you are doing."

"Cape Horn? Of course, we go last year and next year again."

In the cool of evening people gather on deck to watch stars emerge from the twilight—at first single pinpricks and then a swarm, uncountable. The sky tilts steadily night by night, revealing new parts of itself as our changed latitude tips old constellations below the horizon and hoists new ones aloft. Soon there is a thrilling first glimpse of the Southern Cross, its iconic quadrangle pointing toward the antipodes, just below the shadow of Corvus the crow. The North Star sinks lower, steadfast pivot of the heavens until a day at the equator when it will dip to the horizon and vanish. In the northern hemisphere, Polaris will always make an angle with the horizon equal to your latitude—a cosmic geometry first revealed to me in magic diagrams by an astronomy professor, rocketing across the blackboard in a cloud of chalk dust. He had wild brown hair and sideburns like Van Morrison, which we all tried to grow ourselves but of course couldn't.

Among the three thousand objects visible on a clear night are fifty-seven navigational stars, their movements mapped by a line of astronomers tracing back to the Greeks and Egyptians. The names—Al Na'ir, Alphecca, Betelgeuse, Capella—are resoundingly not of today. In the chartroom a GPS plots our path in bright images accurate to within meters, but the ancient skill of celestial wayfinding remains a compulsory piece of what mariners must know. Sailors learn the heavens by rote, memorizing arcs that bind useful objects together. Find the Big Dipper, marked at its tip by the star Alkaid. Trace the curved handle past its end point and across open space to Arcturus, in the constellation Boötes. Continue on to Spica, in Virgo. Somewhere south of the equator in empty longitudes it becomes possible to see beyond Virgo a series of fuzzy blobs, the Magellanic Clouds—in fact not clouds at all but unimaginably dense throngs of stars, dwarf galaxies discrete from our own. The mind explodes.

O N a printed sheet at the chart table are the *standing orders,* fixed instructions by which the watch standers of every seagoing vessel must abide. They differ slightly from ship to ship but share a clear general theme, heavily biased toward calling the captain: One should notify the captain in the event of changing weather, worsening visibility, or deteriorating sea conditions. One should keep clear of all other traffic by a set margin and maintain a sharp watch by radar on all *targets*—the proper if contradictory term in navigation for things one is trying not to hit. The watch officer should of course summon the captain if in doubt about anything, and if the presence of doubt itself is unsure, well, then doubt exists.

Radar measures the range and bearing of distant objects, using a pulse of microwaves sent from a rotating antenna that tracks the timing of their echoes. If a target remains on the same bearing while its range decreases, it will eventually hit you. This is the law of relative motion, well demonstrated in everyday life by the behavior of merging freeway traffic. Imagine someone coming up a ramp to your right, refusing to yield, getting larger and larger in the window until suddenly your door handles are touching. That is a collision. Ship radar has become quite sophisticated in its ability to track other vessels, providing helpful graphics and detailed data on just how close they will get, and when. Despite this, ships still manage to run into one another all the time, for reasons also comparable to life on the highway: Someone was distracted, someone was going too fast, someone was fixated on their plan and simply failed to give way. The ocean is vast and ships are few, yet somehow the first ship you encounter after hours of solitude will, often as not, turn out to be on a collision course, or nearly so.

For all its miracles, radar may fail to reveal everything there is to see of what's around you: perhaps something in another vessel's aspect that reveals its intentions, or the presence of a small boat far from anywhere, visible only by the firefly flicker of its tiny masthead light. Hence the continuing requirement for all ships to maintain that most ancient element of good seamanship, the human lookout: cold, bored, hungry, and not to be multitasked or distracted in any way. Someone standing out in the weather and looking out at what—for 99 percent of the time—is nothing. Somewhere there is a fishing crew still alive because our lookout many years ago saw their unlit panga, late one night amid the shoals off Honduras. Suddenly close enough to see their silhouetted heads in the boat, we altered course to avoid them. Elsewhere in history are the less lucky, those doomed or taken to task according to an inarguable logic: If you hit something, you have failed to keep an adequate lookout. "Res ipsa loquitur," say the lawyers. These things speak for themselves.

In a notebook near the standing orders are the *night orders*, an evolving set of instructions left for when most of the ship's company—including the captain—are resting peacefully in their bunks. Night orders are for the most part routine and will look familiar if you've ever lived with a well-organized person. "Steer a course of 230," I might write. "Let me know if the wind shifts and check in with the lab about towing nets at midnight. Dawn watch should please plan to shoot stars, come and get me if you need an extra hand while this is happening. The cook has asked that we don't open the freezer tonight and advises that brownies are available for late snack. Don't eat more than two."

Above my bunk a porthole shows alternate views of sky and water as the ship rolls to leeward, a more than fair imitation of a front-loading clothes washer. The motion has built to the point where objects cannot be left untended, and an abandoned saucer starts a slow self-powered traverse of the cabin table. I reach to arrest it, one knee hooked under the desktop to keep my chair from sliding. In my world there is still paperwork to do, a tangle of crumpled receipts left on my tiny desk from a week's worth of random purchases in Puerto Vallarta. When we go forth to colonize Mars, the spreadsheet will follow.

After four days at sea the exhausting process of launching a voyage has nearly resolved. The ship environment assumes the feel of a worn overgarment, familiar, if slow and cumbersome to don. It always takes at least this long for me. The stressful departure from home, blurred arc of travel, and sudden arrival in a busy workplace freighted with expectations of competence. The million details and sudden confinement of the ship's routine. Macaroni and cheese for dinner when, in fact, you hate it with a passion. Can't a captain, my friends ask, demand their meal of choice? Perhaps, but at no small expense to goodwill and shared energy. Better to save veto power for some more worthy occasion, whatever that might be. I worried more about this Sisyphean going-to-work experience when I was younger and thought it uniquely my own. Then a revelation, delivered casually in a comment from an old mentor joining the ship to relieve me. Fresh from the airport and sitting, jet-lagged, at a table filled with chatty people happily up to speed with their places on the team, he set down his fork.

"We all hate this part," he said.

⌐——

Our next major task is to make the first in a series of choices about just how we'll proceed to our distant destination. The shortest path between two points on a sphere is a *great circle route*—an arc along the edge of an imaginary disc set to bisect the globe, with your position and destination on its perimeter. Each minute of arc on a great circle is equal to exactly one nautical mile, $\frac{1}{21,600}$ of Earth's circumference. This is how a container ship might plan a journey like ours, or an airplane. Fewest miles, least time, lowest fuel cost. For a sailing vessel the calculus is different. Miles mean little until one considers which way the wind may blow.

Together with the chief scientist I am looking at weather maps, blurry black-and-white facsimiles of the Pacific drawn at sea level. From this vantage the planet appears all water, and no other image says so much in one glance about the collected movements of the atmosphere. Rising air in the ITCZ expands, cools, and releases its clouds of heat and water. From high above the equator it flows back the other way, toward the poles, where dense air hugs the surface and there is lower pressure aloft. The traveling air soon starts to sink, a dry cascade settling from the sky at about 30 degrees latitude, a third of the way to the pole. On our map this process is traceable in the telltale footprint of the North Pacific High, stretched out from Mexico to Hawai'i. It is an ocean-sized pillar of sinking air, with a clockwise pattern of wind at its periphery. Gradients of surface pressure are drawn on the map with *isobars*—lines of equal pressure—4 millibars apart. As on a hiker's topographic map, a steeper slope means closer line spacing, and as a consequence more wind. There is a mirrored twin of the North Pacific High across the equator to our south, in the opposite hemisphere and thus with a counterclockwise rotation of wind at its outer border. Between the two are the trade wind belt and the intertropical convergence zone itself, drawn as an unspectacular pair of meandering lines.

This up-and-down recirculation between tropics and midlatitudes makes up an atmospheric loop called the *Hadley cell*, a semipermanent system that runs the weather at Earth's waistline. Elsewhere patterns are less regular, but still recognizable. Air drifting poleward from the midlatitudes is deflected into a wandering band of westerly winds between 35 and 50 degrees of latitude. Arctic air grows dense and sinks. It pushes south and bends into easterlies. A sharp boundary, the *polar front*, develops between the warmer westerlies and the cold polar flow, spawning turbulent eddies that become the storms of the temperate regions.

Here in boreal fall the ITCZ is near 9 degrees north latitude, still well above the true equator. How we choose to cut across it will affect our progress going forward. Most ships begin by looking for a break in the line of convection, a sort of mountain pass that will give the hope of a minimal interruption to settled sailing. In our particular case, we'd also like to avoid crossing too far east, where an upper quadrant of the South Pacific High is bending winds in an unfavorable direction. And as our ship is in its full appellation the *research vessel Robert C. Seamans*, there is specific interest in the water itself—some parts of which are more compelling than others.

Of particular note on this voyage is the *cold tongue*, a transient mid-ocean lake brought to the surface by currents near Peru. In years of brisk trade winds this feature spreads far offshore, forming a startling temperature anomaly in the otherwise warm tropics. Nutrients delivered by the cold tongue feed an explosion of organisms, sustenance for entire food webs and their coupled economies: krill, anchovies, squid, and tuna; Peru, Chile, Panama, and Ecuador.

The chief scientist pulls her hair back and leans in for a closer look. Perhaps the smartest person I know on a first-name basis, Dr. Kara Lavender is a physical oceanographer, her bias tilted away from the living world and toward the ocean itself—a dense liquid counterpart to the atmosphere with its own epic overturnings of energy and dissolved compounds. Kara's appetite for raw data is not one to be satisfied by the mere random swimmings of biology. Only equations will do.

There's a lot going on where we are headed. Westbound ocean currents straddle the ITCZ, driven by the trade winds in their respective hemispheres. Near the axis of the convergence a countercurrent brings water back east—and this year the cold tongue is highly active, its saw-toothed pattern of swirls visible on infrared satellite images downloaded before our departure. A thermodynamic symphony is being played in the ocean. I look at the maps and try to imagine the surface as we will see it—the mixing of warm and cold, the piles of cloud, whirling seabirds a telltale of nearby life. Kara points with her pencil.

"That's an interesting spot. We'd find things to look at there, I'm sure."

"Mm."

I am unsuccessful at hiding my indifference.

"Is it too far east?"

"Farther than I'd like to be. The winds may be a bit more out of the south than we'd prefer in that area, but it's not impossible."

On research vessels the captain and chief scientist have joint interests in where the ship goes, sometimes shared, at others divergent. There are negotiations. Aboard educational ships like this one it's uncommon to encounter real friction, but at higher levels careers may be at stake. Egos, grant money, and the pressures wrought by reaching through small windows of opportunity into a dynamic environment. Oceanography is a field starved for data, and much rides on the fleeting moment when a ship at sea is in the right place with the right instruments. I have had a few lively discussions over time, rarely true disagreement. To me the relationship is mostly one of enrichment. How interesting to have the solitude of command buffered by interaction with a great nearby mind, hard at work on problems that are fascinating, if not purely nautical. FitzRoy had Darwin, after all. James Cook had Joseph Banks, and Captain Jack Aubrey had Dr. Maturin.

Not everyone agrees. Once, in giving a tour to a colleague from another organization, we finished with coffee in my cabin.

"And whose bunk is that over there?"

"The chief scientist."

"You're joking. The captain and the chief scientist share a cabin?"

"That's right."

"But . . . that means you must have to speak with one another."

❧

O N the morning of December 7, the trades become fitful, and the sky turns its palette to gray. Near noon there is a torrential round of rain without much wind, and by evening we are floating past great vaulted pillars of cloud, lit sideways by the sunset. The wind is still easterly but very light, and with many miles left to go we start our engine, keenly aware of an advantage that our sailing forebears did not have. Without wind we find ourselves lurching over a tall swell driven from the north by some distant storm, a reminder of violence elsewhere even as we drift through the tropics. The following night is unsettled, dense bands of rain interrupted with splashes of clear sky. I'm called frequently by the crew, mostly to assess the severity of passing squalls, but at least once just to look. Near midnight it is the mate, holding back my curtain to deliver a non-emergent summons:

"Cap. All is well, but you've got to see this."

On deck a towering cell of cumulonimbus is passing astern, between us and the nearly full moon. From underneath it a torrential cascade of rain, an illuminated waterfall six thousand feet high. The assembled watch have paused in their business and are gathered at the rail, just staring.

I have a favorite book, *N by E*, in which the author—artist Rockwell Kent—tells the story of being shipwrecked on a trip to Greenland in 1929. Forced to walk over the backcountry for rescue, he and his mates come upon the twilit vision of a glacial stream, pouring from great height into a valley of stunning beauty.

"Maybe we have lived," he says, "just to be here now."

⌁

OUR crossing of the eventful doldrums takes most of another day, ending abruptly with a dry puff of air at the start of the dawn watch. Lines of cloud recede behind the sudden full expanse of constellations. Occasionally the stars vanish amid residual lightning flashes that somehow find their way across the newly clear sky. A sailing ship under engine power is a restless conveyance, where a hundred things meant to be under tension are suddenly free to crash about unchecked. Slack lines beat a tattoo while the diesel roars out its demand for fuel and attention. The relief is thus palpable as the trade winds reappear, now from the southeast and startlingly cool. Sails are hoisted with a willful glee. We are at the fulcrum of our passage, with 1400 miles behind us and 1400 still to go. It is an important, if intangible, benchmark—our exotic goal still just a verdant dot in the imagination, like some outer planet in a solar system only partially traversed.

Over the next day it grows cooler still, the equatorial mornings decidedly nontropical in their feel—filled with damp gray mist, otherwise empty. We are a thousand miles from any land, in the most international of international waters. It has not so far been a busy trip for the radar. Since leaving Puerto Vallarta, we've seen three ships, no hint of any other solid object. At breakfast the startled lookout suddenly reports a boat nearby, a rusty white-hulled vessel with an elegantly flared bow. Perhaps a second one is visible floating in the distance, but the low haze and long deep swell make it hard to be sure. There is a sudden brief burst of jabber on the radio in a language I don't recognize, and then silence. Ships are expected to communicate with one another in English when discussing navigational issues, but if these two are partners in some distant-waters fishing fleet, they could be speaking Ukrainian, or Basque. In an hour they are gone.

Our own nets fill with fascinating creatures—squid, jellies, tiny lantern fish from the Myctophidae family, and some snaggle-toothed beast called a snake mackerel, clearly thwarted in pursuit of its lunch. We deploy other instruments to great depth on the winch wire. There's an ingenious device called the *rosette*, a cluster of tubular bottles set to close themselves in series and return with precious slugs of water from each submerged stratum. It resembles a rocket launcher. We are like a space probe, collecting clues one bottle at a time from the midst of an unmeasured expanse. The science crew gathers around their equipment, sorting samples and reading from gridded sheets of data. Some are wearing ski hats.

"I thought it was supposed to be warm near the equator," someone complains.

"You've been licked by the cold tongue," says Kara. "Embrace it."

3

THE CLOUD FOREST

A hundred miles north of the equator we are overtaken by a strong squall on the evening watch. I've just set my book aside and drifted off to sleep when an apparition appears by the glow of the cabin light, a trainee sent by the mate to tell me that a line of dense cloud has appeared upwind of us, and that it might be time to take in sail.

It is very dark on deck when I get there, the air heavy and soft, almost without temperature. The wind has picked up and the ship heels with a concerted momentum that reminds me oddly of skiing, those instants where you point your tips downhill and are suddenly aware of all that gravity can do. As I peer around to identify faces, a flash of lightning illuminates the sky behind us, exposing a steep wall of cloud with ragged spires peeling off it far above. The world, if possible, looks even darker underneath. A seabird of some sort, perhaps a booby, has been attracted by our lights and is pacing us to windward, his elongated shadow flickering across the mainsail like a zoetrope animation. Somebody speaks.

"Hi, Cap—looks like maybe some excitement on the way. What do you think? I've got my people standing by to take in the topsail, and the main halyard is ready to run."

It is Jeremy, who from his tone could be offering me toast at the breakfast table.

Sailors rarely spend time debating in situations like this. I take the wheel as he leads the watch forward to take in sail, individuals vanishing to their preordained stations in the blackness—lines with exotic names that they must now find in the dark, the halyard coils prepared specially to avoid the potential disaster of fouling. There is shouted communication, a flurry of activity, and just as the work concludes an abrupt wall of wind-driven rain, heeling the ship over another seven or eight degrees—even with half of her canvas now safely furled. I steer to keep the wind astern, aware of a sudden flattening of the sea around us, pounded down by the torrent and a heavy draft of cold air. Small tempests of spray spin up from broken wave tops. Somebody asks me how hard I think it is blowing, and I'm not sure, though afterwards we'll see that the anemometer has recorded a peak gust of 52 knots, a bit more than double what we'd started with. As the watch comes aft, there is another flash of lightning, the clap of thunder just above us like an artillery shell.

Fifteen minutes into the maelstrom the wind abates, and stars reemerge as the squall line tails away to leeward. A broken coffee mug rolls about disconsolately in the scuppers with some other random debris. My glasses are useless, and I have water trickling down the neck of my raincoat and onto my T-shirt. The watch regroups and begins to plan for the substantial task of setting sail again—they will do this with the oncoming team in an hour, shaking out wet gear while forks of lightning play dramatically on the horizon.

Thunderstorms are supreme expressions of the energy involved when water and heat make their way around the atmosphere. They begin with convection, frequent in the tropics as an abundance of warm moist air grows buoyant and rises aloft into colder strata. Clouds formed by this process are in the cumulus family— some angry, as we've just seen, but others benign, fleecy columns formed on fair days as the sun makes patches of warmth near the surface. Whether these cheerful juveniles rise to a threatening maturity depends on the particulars. The real engine for cloud growth is driven by what forecasters call *instability*, the relative density difference between stacked parcels of air. If warm air near the surface floats up into a place that is sharply cooler, it becomes more buoyant and rises even faster. Results hinge on the contrasts. A cumulus cloud expanding upward through cold dry air with an abundant supply of warmth and moisture below it becomes an atmospheric chimney. Air is inhaled at the bottom and whisked aloft, its buoyancy boosted by periodic injections of heat from condensation. Rain cascades from on high, pulling cold air down behind it. The friction between molecules strips away electrons, creating static potentials that discharge as lightning.

The expanding cloud tower vaults upward until high-level winds shear it off into a parapet, its vertical progress damped by interaction with the stable stratosphere. Converging warm air and outflowing cold spread the cloud base out like the foot of a giant mollusk inching across the landscape. This is a *cumulonimbus* cloud, the suffix *-nimbus* referring to rain. Generically, it might be called a thunderhead; a squall, if you're a mariner; or a *tormenta*, if you are speaking Spanish. A tormenta can be merely the adult stage of some ambitious afternoon rain cloud—or an internal player in some much larger event. Cold fronts sweep long lines of thunderstorms across the landscape like seismic theater curtains. Tropical waves, troughs, monsoons, and cyclones all inhale hot moisture at the surface and whip it aloft in great expanding billows. Meteorologists watch these eruptions from satellites and judge their height using infrared imagery—the higher a cloud goes, the colder it gets, with the coldest spots on the film representing the tallest clouds of all, places where convective energy is at its peak.

Regardless of what you call them, thunderstorms deserve caution for the force that they are capable of releasing onto the surface. In February of 2010 the tall ship *Concordia* was overtaken by a squall off Brazil, and in the process knocked over by a gust until she flooded through her hatch openings and sank to the bottom. The crew all escaped into rafts and were rescued by a pair of merchant ships thirty-six hours later. Debate remains over just what went wrong that day. An official report issued eighteen months later by the Transport Safety Board of Canada concluded that the vessel had not prepared sufficiently in anticipation of what were, in the board's view, hazardous but routine conditions: During a period of expected squall activity, they'd had too much sail set, too few people ready to react, and too many hatches open. A colleague of mine was the captain, and he describes the vessel being struck not by a routine squall but by a *microburst*—a sudden blast of wind blowing not sideways, but *down*, at more than a forty-five-degree angle to the surface. This meant that his ship was subject to an increase in heeling force as she leaned over, getting pushed still further by a gust blowing nearly at right angles to her tilted sailplane.

I'd seen something like this once myself, west of the Azores on a fall crossing of the Atlantic. I was the second mate on that voyage, engaged in some sailorly task aloft when a dark cloud passed above us, and I saw a line of whirling spray come churning across the water in our direction. Seconds later we were hit by a banshee of cold air that seemed to come from everywhere at once. We were still carrying our largest squaresail, and I watched its sheets all part together with a bang as it transformed itself into a huge flailing flag, miraculously subdued before it could self-destruct. I recall leaping for a taut line and sliding down out of the rig like I was Jack Sparrow, with the sort of idiotic and once-celebrated heroism that you got to partake of if you were twenty-five at the right time.

The origins of a microburst are not hard to grasp if you can visualize the anatomy of a thunderstorm. Warm air is whisked vertically aloft, while rain-chilled cold air plummets back to earth and splatters out in a great fan behind. Depending upon your location relative to the splatter zone, you might see a sudden wind increase from almost any direction and orientation—horizontal to nearly straight down. Not all thunderstorms generate powerful downdrafts, and not all downdrafts are powerful enough to qualify as microbursts, but the effects when they happen can be dramatic. The US National Weather Service explains how a mass of cold air and water may hang trapped high in a cloud until updrafts weaken and the cold portion drops out all at once like a bomb. There's a video on their website that shows what looks like a giant bowling ball of wind and rain—perhaps a mile across—falling out of a cloud in slow motion and exploding onto the surface. Something like this happened in 2007 near my home in Maine, when twenty minutes of utter mayhem turned the scenic village of Castine into a war zone. Century-old elm trees were broken into jackstraws, roofs collapsed, and a local preserve of spruce woods was thrown down as if by the hand of God. I'd seen damage like this before on hilltops in Newfoundland and wondered at the cause. Now I knew.

Pilots face a major threat from these events. Commercial aircraft today are required to carry Doppler radar equipment, specialized for detecting the wind shear associated with thunderstorms. Much of this innovation was driven by the loss in 1985 of Delta

Flight 191, a Lockheed L-1011 that crashed on approach to Dallas after flying through what was later determined to have been a microburst. First a sudden tail wind robbed the wide-body jet of its airspeed; then a sinking blast from above forced it onto the ground just a mile short of the runway. Thanks to a successful confluence of action and innovation, events like this are now rare at modern commercial airports, where flight controllers use multiple radar inputs to identify the dropped-milkshake wind signature of microbursts and warn approaching flights accordingly. Recently at a conference, among a roomful of pilots and aviation meteorologists, I learned about how the number of deaths linked to microbursts has shrunk to near zero over the last three decades. A triumph of public safety, equal to the air bag.

My friend Michael flies jets for FedEx and says that it is now routine to carry extra fuel along with his packages when it looks like dodging wind events might be necessary.

"Summer thunderstorms rolling across the Midwest pretty much make this a daily issue on launches into and out of DFW, IAH, ORD, MEM, et cetera . . ." he explains, the airport acronyms rattling out like familiar street names. "The same can be said throughout the year for ABQ, SLC, and DEN, as they are close to mountains which the air masses are flowing over and eddying around. But even the longest flights only have to look ahead fourteen to sixteen hours, so we will still dispatch to a known area of activity but bring along extra gas to give us options— alternate routing, a few turns in a holding pattern, or to just get the hell out of there and divert to an alternate field."

Blessed with speedy craft and the miracle of Doppler radar, pilots can delay departures, redo approaches, or simply go elsewhere. Slower than the weather and with simpler equipment, ship crews are more obliged to recognize what's coming and be ready when it arrives. Marine radars lack the direct ability to measure wind, though they can still see rain at the surface—a proxy of sorts for what's happening in the atmosphere above. Fast-moving bands of rain may indicate the presence of stronger winds—and sharp-edged outlines in the echo may warn of similar dangers. Frontal boundaries acquire a linear aspect on the screen, while a sudden clear gap in a dense band of rain might indicate a blast of descending cold air. Or it could be nothing at all. It's an inexact science. Sometimes your best bet is to just to look around and react to what you see. A bulge descending from the cloud base, called a *mammatus*, can be one potential sign of impending bombardment. Never forget what is happening outside the window, I tell the crew. Nonetheless, on a dark rainy night, what your eyes see is nothing. You can look out the window all you want, but it will tell you very little.

Power-driven vessels carry a greater tolerance for short sharp shocks of sudden wind, but sailing ships, set up by design to collect it, must be more careful. The brigantine *Albatros*, once owned by the writer and adventurer Ernest K. Gann, was knocked down and sunk in the Gulf of Mexico by a microburst in an event commemorated (if not perfectly retold) in the movie *White Squall*. The American schooner *Pride of Baltimore* sank north of Puerto Rico in 1986 with four lives lost, in a case with some similarities to the loss of *Concordia*—including the circumstance of a strong frontal system passing across warm tropical waters near a continental margin.

Most who sail treat squalls and thunderstorms with an abundance of caution, which is not to say that the crews of these vessels weren't doing the same. I shipped once with a Navy officer who worked on a nuclear submarine. He talked about the drills they had for "scramming" the reactor in emergencies—a technical process I never quite grasped, except to think that it had a perfect analogue in my own wind-powered world: Kill the power before it kills you. Strike all the sails you have, as fast as you can, starting with the largest. That is the catechism, anyway. As I'm sure is the case with nuclear reactors, there are nuances particular to the situation. Usually the big sails come down first, or perhaps just after you've taken down the ones set farther aloft, which exert greater heeling leverage. People with small-boat experience might have the habit of rounding up into the wind to take in canvas, but on a large ship offshore you frequently do just the opposite, steering to keep the wind somewhere behind you. This is called *bearing off*—with the wind at your back things become apparently calmer, and the boat stands more upright. This virtual lull may buy the time needed to get people organized and bring in canvas. Some squalls look less threatening than others, and perhaps in those cases you'll stop short of shutting down the whole reactor and leave a few sails where they are.

Sail drill, or the process of making the sails do what you want when you need them to, is at the very heart of operating a ship safely at sea. Modern designers have worked hard to engineer greater safety and convenience into sailing rigs, but aboard traditional vessels everything requires a carefully coordinated effort, the hauling and easing of multiple lines together while powerful forces are brought under control. Ships are most vulnerable early in voyages, when the crew are unfamiliar with the equipment and new at working together. It is comparable to the first scrimmages of a new soccer season, with much more serious consequences. As with any team effort there are usually a few people who really know what's going on, and a cadre of followers—waiting for instructions but otherwise limited by inexperience, anxiety, seasickness, or hypothermia.

While sailors are often asked to share dramatic stories of storms at sea, most weather events are in truth comprised primarily of tedium and endurance—as with the mountaineer huddling in their snow cave while the blizzard blows past, what's left to tell is mostly of waiting. The real benchmarks of memory are often the sailhandling evolutions themselves, punctuations of excitement that might stand out sharply long afterwards. There is the green hand racing frantically into the darkness to find a line they learned about just days before, sending exactly the wrong sail crashing to the deck. A topsail in a gale, stuck halfway down for unknown reasons and flogging itself to bits, fifty feet in the air. The main boom, flagpole-sized, broken in two by an accidental jibe and now swinging in space like an overlarge martial arts weapon.

For all of these occasions, there are procedures that can to some extent be superimposed over the chaos. On my ship inbound for Cádiz one drizzly morning there was suddenly a line of waterspouts, winding down out of the cloud layer just a quarter mile to starboard. A form of marine mini-tornado, waterspouts can hatch under developing cumulus clouds where convection is enhanced and there is a source of turbulence—or *vorticity*—at the surface. The fast-building cloud sucks air up from underneath itself, and the converging column whirls into a tightening spiral like a spinning figure skater. Estimates of wind strength in waterspouts place them near the bottom of the five-step scale used for tornadoes, with gusts as high as 75 or 80 knots. Combined with their generally small diameter, fifty or a hundred yards, this makes them the analogue of a pint-sized hurricane. I don't remember how much sail we had set. The answer was too much. The new chief mate had a plan. Better to say, they made one from scratch. A dozen idle hands ran up from breakfast and heard exactly where to go. Our spread of wet canvas folded like a book, and we were ready in time to watch the rotary fracas of wind and water pass just astern—the closest look I ever want to have at such a thing. Clad head to toe in raingear, we were wet to the skin.

NINETY minutes after my unplanned wake-up, the combined force of two watches is finishing the task of setting sail again, restoring in half an hour a configuration that, with the help of adrenaline, took only minutes to strike. To our west the departing squall line returns to view at odd intervals, picked out by sheets of white light and occasional peals of thunder, long delayed. Lightning is a mega-scale version of the shock you get from sliding across the carpet and touching a doorknob. As you shuffle your feet, electrons are sloughed from one molecule and added to another. A difference in potential develops, equalized on contact by a tiny spark. The cumulonimbus cloud with its racing columns of air and water is a veritable power plant of static electricity, the numbers involved so large as to be meaningless. A single bolt of lightning might last a tenth of a second but holds enough energy to supply your house for a week—all delivered at an impractical thirty million volts. An active storm might deliver a thousand flashes in a minute, slender arcs with a temperature near 50,000 degrees Fahrenheit. The abrupt heating of the nearby atmosphere creates pressure waves that we hear as thunder.

Depending on the arrangement of differing charge potentials, lightning can strike the ground, other cloud masses, or even another place in the same cloud column. The most awesome display of this I've ever seen took place in the Gulf of Mexico late one winter, when a stalled front parked a long ribbon of warm wet air above us for what felt like a week. Each evening in the cooling sky there'd be an immense ragged blossoming of cumulonimbus clouds overhead, stretched into a loose line along the remnants of the frontal boundary. It was calm and dry at the surface, but above us the lightning played all night like welding flashes, bright enough to force your eyes closed.

Lightning is most common in environments where warm air can rise rapidly with a plentiful source of moisture nearby. Florida is the thunderstorm capital of North America, as powerful convection over the mainland draws warm maritime air steadily in from both sides of the peninsula. Venezuela is world champion, its supremacy owed to a confluence of moist Caribbean winds and cool mountain air around Lake Maracaibo, a mixture that creates lightning at a rate approaching three hundred days a year. When lighting strikes, it's attracted to tall solitary objects, where positive charges accumulate in response to negative leaders reaching out from the cloud base. The Empire State Building is famously hit about twenty-five times a year. Like other, similar structures, the building goes generally unharmed, as the current is carried to ground through a series of dedicated conductors. Vessels at sea require the same sort of protection, with the metallic elements in their rigging connected to a grounding surface in the hull. This happens somewhat automatically with steel ships, but boats built of wood or fiberglass must take steps to ensure that their rigging is properly protected. After Ben Franklin's famous kite-and-key experiment showed that lightning was in fact electricity, sailors started to trail lengths of chain overboard for this purpose. In modern times it's typically done with a length of stout wire led from the rigging to something heavy and metal underwater.

Even when dispersed through a ground, the extreme current of lightning can generate a damaging electromagnetic pulse, a very miniature version of what you hear about in connection with nuclear explosions. This creates a certain all-bets-are-off condition where electronic equipment is concerned. When my ship was struck once while docked in Barcelona, everything survived except for an antenna tuner, housed in a box far up the mainmast. It all looked OK, but deep in its circuitry (visible only to the technician) was a tiny capacitor, fried to a crisp. Once, on another boat,

I awoke to a horrific thunderclap and found the masthead shattered into steaming fragments, scattered about the quarterdeck. Also ruined were a battery charger and our radio, which made happy static but never allowed us to speak to another vessel again.

Lightning can do catastrophic damage to trees, as the sudden blast of heat causes the sap to flash into an explosion of steam. The interval of heat is usually too brief to start fires directly, but it's not uncommon for wooden structures to burn after being ignited by smoldering electrical equipment inside their furnishings. Thousands of unlucky people are struck each year. Many of them survive, but they frequently suffer injuries to internal tissue from the electromagnetic energy. Most vulnerable are those standing exposed in open areas, like a mountain ridgeline or soccer pitch. I learned in a wilderness medicine class that it's possible to get a shock from current traveling through the ground as well as by a direct strike. Hikers who can't find shelter are sometimes taught to sit on their packs for insulation, something I've never had to do but which must feel like being marooned on the smallest, loneliest island on Earth.

In my travels by air, I always pause to glance with admiration into the cockpit, where pilots are calmly preparing the tools of their awesome trade. They are ready to meet whatever comes in their crisp shirtsleeves, never needing to begin by climbing into wet oilskins and fumbling for boots in the darkness. I pass by like some primitive professional forebear and wonder how this must feel for them. They have no idea.

✑

THERE's a lot going on inside a cloud, most of it poorly understood by the average person. Or fairer to say, it's not a priority for most people to understand. I spent some time in graduate school researching this and learned that many students thought that clouds were made only of water vapor. This is a small but important misconception, since water vapor is invisible. Clouds are in fact clusters of water droplets and ice crystals spawned by condensation or *deposition*, the process whereby water vapor converts directly to ice. These changes in state are meaningful, since the cycling of water is a primary part of how heat gets distributed through the atmosphere. Heat is a quantitative measure of *energy*, which to physicists means the capacity to do work. The standard unit for energy is called the *joule*, somewhat obscure outside of science. One joule is equal to one kilogram-meter squared per second squared. Exactly. More familiar to most may be the dietary calorie and its smaller cousin, the thermo-calorie. Adding heat to something will change its temperature at a certain rate, depending on the material. One thermo-calorie will raise one gram of water 1 degree Celsius. One dietary calorie is a thousand thermo-calories, meaning that your Snickers bar easily holds enough energy to boil a liter of water.

The relationship between heat and temperature is linear, as long as there is no phase change in the material. Heating water is a good example: A gram of liquid water requires one thermo-calorie to get 1 degree warmer, but as the water begins to evaporate, each gram needs an additional 539 thermo-calories to convert itself into its more energetic gaseous phase. This is what's known as *latent heat*, since it is energy stored in the movement of the molecules but not measurable as a temperature change. In this way, water vapor serves as a reservoir for energy in the atmosphere, holding latent heat until it is eventually released during the condensation process. The energies involved are considerable. A kilogram of water vapor releases 540,000 calories of heat when it condenses into raindrops. A good rainstorm over one square mile of ground might process sixty million kilograms of water in an afternoon, yielding enough energy to keep a medium-sized town going for a day.

MOST clouds don't drop rain or snow onto the ground, because their water particles are too small to survive the trip. What's actually happening is an equilibrium, where droplets fall out and evaporate, only to be replaced by fresh condensation and deposition in the cloud mass. Clouds transform in this way before our eyes, as their visible parts emerge and vanish. Eventually the ice and liquid may aggregate into particles large enough to reach the surface, a joining that takes place in two distinct ways: *Collision and coalescence*—which is just what it sounds like—involves liquid bits banging into one another until they form a blob large enough to qualify as a raindrop, generally at least 0.2 millimeters—a hundredfold increase in comparison to the average size of a cloud droplet. In colder clouds, the hooking up of particles is dominated by something called the *Bergeron process*, an elegant mechanism whereby water evaporates from liquid droplets and deposits itself onto nearby ice crystals. The latter might reach the ground as snow or melt on the way and fall as rain.

The important thing to know is that most clouds are a melee of mixed ice and liquid water, depending on the specific temperatures and humidity involved. It also helps to recognize that not all precipitation starts its journey in the same state in which it arrives. All sorts of stuff happens. Snow may fall through a layer of warm air and arrive as rain at the surface. Rain may fall through cold strata and land as sleet. Hail is made in thunderstorms, when updrafts send rain into the high parts of the cloud, where they freeze into icy projectiles before falling back toward earth. Hailstones might make several such round trips in the cloud column, growing a bit on each circuit until they are large enough to dent roofs and break windshields. In a bit of droll genius, the National Weather Service correlates the size of hail with common objects—starting with peas and moving on through golf balls and teacups before ending at grapefruit. The largest hailstone ever recorded weighed two pounds and was the size of a clock radio.

Freezing rain—a fascinating if inconvenient phenomenon—is produced when ice precipitation melts on its way through a warm layer before getting re-chilled near the ground. Through a process known as *supercooling*, the liquid drops may be cooled below their freezing point, but don't actually solidify until they strike a surface. Freezing rain is typically a fleeting event, but in the right conditions it can last long enough to deposit a damaging load of ice onto trees and structures. This happened in the northeastern US in January of 1998, when a layer of warm moist air made its way up the Mississippi valley and got trapped above a slice of frigid air pushing down from Canada. The resultant ice storm did $7 billion in damage and created more than four million power outages across Quebec, New York, and New England. Electrical line towers bent like licorice under their frozen load. I was visiting my in-laws in Florida at the time, but my father was happy to tell me all about it later as we sat around his woodstove waiting for the power to come back on. It took weeks to repair the damage. In town the hotel parking lots were filled until March with utility trucks and their crews, some from as far away as Texas.

❧

ONE day at lunch I walk with my wife down the long meadow across the road from our house. It is early spring, mud season, and the ground squelches with water trapped in clay above the frost layer. Drifts of granular snow recede to show patches of bare ground like the seabed at low tide. I am leaving for sea again soon, after two blissful months at home with the coffeepot and a thousand small projects. This is the true and hidden profession of most mariners.

"The truck is going to need new summer tires this year," I tell Karen.

She steps blithely through a puddle of slush and turns to answer, her feet invincible in old bunker boots. This litany of action items is as much for my benefit as hers, a check box in my recurrent abandonment of our living situation.

"Now *that's* exciting."

"Yup. I can leave my checkbook."

"Yay! It will all go for shoes."

"Of course. Don't forget to eat."

"Roger that. Here's wishing you only type-one fun on your trip."

This being the category of events that are fun while they are happening, not when they are over. A good movie versus bungee jumping, for example.

"Thank you."

It's a clear blue day and the airplanes passing overhead draw long lines with their exhausts, bending slowly as they turn toward the large coastal airports to our south. These trails are a form of *cirrus* cloud, plumes of ice crystals deposited by water vapor from the jets' engines as they skim across the top of the troposphere. They are often a harbinger of rain, indicating that the air is near saturation and thus primed for a transition to clouds and precipitation. Moisture ejected from cyclones and swept along by upper winds will also spread out into a thin fan of cirrus, a slight blurring of the outer sky which is the first indicator that a change might be coming. Mare's tails, these clouds are sometimes called, in part of an old sailor's adage that advocates for caution when they appear. Cirrus are the highest clouds around. They start at around 16,000 feet and may form as high up as ten miles. Thin filmy things, they maintain the appearance of a feathery scrim more than an opaque layer. Hazy rings of cirrus may loom around the sun or moon when weather is imminent, refracted halos that appear before any hint of cloud is evident to the eye.

A warm front forms when a sheet of warm air slides gradually over a mass of cold, producing a blanket of cloud at their interface like jelly spread between two slices of bread. This is a *stratus cloud*, the product of a temperature transition between two overlapping layers. Stratus clouds are dreary to behold and easy to identify—if you walk outside and see nothing but a featureless layer of gray, you are observing stratus clouds. Think Seattle or Copenhagen. They are connected with stable and uninspiring conditions, often bring steady rain or snow, and may gradually lower and thicken if the feature that's producing them is headed your way.

Clouds with bases above 6500 feet are considered mid-level clouds, indicated with the prefix *alto-*, which then combines with the appropriate suffix to indicate whether a cloud is the product of convection (altocumulus) or frontal lifting and layering (altostratus). There can be stratocumulus also, hybrid flat-topped puffy clouds that form when convection stalls against a stable layer of air above it. Fog is a cloud too, technically also in the stratus family. Maine is famous for fog, particularly in the warm summer months. It's also famous for cold water, and there's a connection. Much of New England's summer air comes from the subtropics, carried along by the southerly winds at the edge of the Bermuda-Azores High and moistened by the warm waters of the Gulf Stream. The Stream heads off for Europe at a point somewhere south of Cape Cod, but the warm air keeps on going, north and east across the cold outflow of arctic water that fills the Gulf of Maine.

Warm air that's been parked over water is often carrying just about all the vapor it can hold. Forecasters would describe this as a case of high *relative humidity*—a value that measures how much moisture the air contains against how much it can theoretically transport: 50 percent means half full; 100 percent means that the air is saturated, or at full capacity. Chilling an air mass below a certain temperature will drive its relative humidity up past 100 percent, forcing water to precipitate as condensation. The specific value of this condensation temperature is called the *dew point*. Pilots pay close attention to this, as it is an indicator of when clouds will begin to form. Dry conditions mean a lower dew point temperature, since dry air can get much colder than wet air before it saturates. In cases where air is cooled by lifting, clouds appear at the instant when things get cold enough to trigger condensation. This is a uniform point for any given air mass, and the reason why clouds are so often flat on the bottom. In meteorological terms this is called the *lifted condensation level.*

Condensation typically needs a solid surface to get started on, and the seeds of clouds are microscopic particles of dust, soot, salt, and clay that drift about the troposphere. Very clean air, with an unusually low level of particulates, can actually hold vapor well past its theoretical saturation point—a phenomenon that has led to experiments with artificial cloud seeding, where aircraft drop fine crystals of dry ice or silver iodide into clouds to induce rain. This sometimes works, though after fifty years of testing it's still not clear among scientists whether the process can be counted on to yield a significant overall difference in precipitation.

If we go back to New England, consider a summer current of southern air making its way north across the Gulf of Maine. The air is warm and packed with evaporated moisture from the mild mid-Atlantic, but in passing over much colder water it gets cooled rapidly below its dew point. The result is fog, a low mass of cloud droplets dense enough to block out all evidence of the surrounding world. Navigating in fog can be a frightening experience, particularly for the novice—but its qualities give a good demonstration of the interplay between moisture and heat that drives cloud dynamics.

My first prolonged exposure to fog came aboard passenger schooners in Penobscot Bay, where often as not we would leave port on a Monday and sail off into an opaque wall making its way like a river of poured cream up the estuary. Sometimes the boat ahead of us would vanish with its flags still flying above the cloud layer before we were swallowed up ourselves. The captain, supreme in his confidence of where things were, used only dead reckoning to navigate—a wristwatch, a few initial bearings, and a carefully steered course toward a place he knew he could find. After a vertiginous interval of groping through a bubble of watery light and odd refracted sounds, we'd be told to look out for something, and a buoy or headland would come lurching out of the murk, instantly close at hand. From this captain I learned that it is common to find clear air in the lee of an island, where the sun-warmed terrain will burn away a small patch of the mist. Be careful late in the afternoon, he admonished, when things cool off and the fog can come back much faster than you expect. Sparing in his admiration for others, he had little sympathy for those who ran into trouble trying to navigate too precisely when running blind. Aim at stuff you can afford to miss, he told me. This was a sound, if imperfect, strategy. During my last year aboard we ghosted one afternoon up to an indistinct shoreline, past a navigational buoy just slightly too far away to identify. I was sent away in a boat to read its number and found that the buoy was in fact a big man in green waders, with a clam rake.

"Where do ya think *you're* goin'?" he said.

I shrugged and turned around, the schooner now swallowed by the mist. I took my best guess and set out, 90 degrees off course until I heard the lunch bell ring and made a correction.

Maine gets fog in the winter also, through a reversal so extreme it makes the cold ocean look warm. When bone-dry arctic air makes its way out over the water in the aftermath of a cold front, the temperature differential between air and sea might be 50 degrees, enough to make the bay steam like a bathtub. This is *sea smoke*, where moisture above a warm liquid surface cools and condenses briefly before evaporating again into the dry surrounding air. You hear less about sea smoke than summer fog, but for those obliged to work on the water during winter, it can be a real challenge. Often there is *rime ice* too, a glaze that forms as the supercooled fog droplets freeze onto surfaces.

The late Matinicus Island fisherman Dick Ames told me about trying to find his way home in his lobster boat one smoky January morning with his pilothouse windshield iced over. He could see by poking his head out his side window, but it was just too cold to do that for more than a minute or two at a time. He'd just pulled his head back inside when BAM! his vessel struck a crashing blow against what turned out to be a neighbor's boat moored near the tiny harbor entrance.

"Then it was no problem," he said. "After that I knew just where I was."

4

THE SERPENT'S COIL

t's hot and sunny now on deck at midday, enough to drive you into the shade if you've got a choice. The trade winds have returned, steady from just south of east, and the ship slides along as if on a rail. There are dry starry nights, the evenings electric, with horizons the color of watermelon rind. Orion, recumbent, loops overhead in a great arc. We cross the equator near 132 degrees west longitude, just after midnight on December 17. North along our meridian the next bit of land is British Columbia. South is Antarctica. The latitude display on our GPS reads, briefly and thrillingly, 00° 00.000'.

Sailors who have crossed the equator are known as *shellbacks*. It's an honor historically bestowed only after a certain rite of passage, typically involving an appearance by King Neptune, a short haircut, and a round of abuse at the hands of the previously initiated. For our group, a modern aversion to hazing competes with the desire of individuals to feel like they've achieved some distinct status by their passage. There will be no tar and feathers on this crossing, we agree, or rotten-egg omelets. When Neptune appears, he bears a suspicious resemblance to the chief scientist. The inductees run a gauntlet of tempered reprimands, wet fishing nets and fire hose baths, and a ceremonial dinner of cold oatmeal, all met with light resistance from rogue elements in the engineering department. Haircuts are not offered, but the students get hold of some clippers and barber themselves anyway. In short order the whole crew look like fans at a Ramones show in 1977.

Afterwards there is a short meeting in my cabin to discuss details for tomorrow, a day whose routine will be altered by an all-out cleaning of the ship. "Field Day" is our name for these weekly exfoliations, a title I took to be our own until someone informed me that it was in fact a military term, borrowed from a time when infantry units would clean their gear and line up outside for inspection. We make our plans amid steady interruptions from a scientist who's sitting nearby with his laptop, scrolling through microscope images of captured plankton. The things he sees in this miniature world are too compelling not to share.

"Look at this," he says.

Caught in the translucent glow of the screen floats something that would haunt your dreams if it were more than twelve millimeters long. The phronimid amphipod is a planktonic bad citizen that lives by invading a family of tubular jellies known as salps— burrowing in and co-opting the hollow husks as nurseries, where their larvae grow in clutches before bursting forth to begin the cycle anew. Hatchet-headed, set amid their pulsing mass of young in the emptied shroud of the host, these animals are said to have been the inspiration for Ridley Scott's *Alien*. Nobody seems fully sure, but there is great comfort in knowing that they are in earthly life about the size of a pencil eraser.

The crew gather their notes and stand to leave, as the scientist flicks through to another slide and again rotates his computer to display one last image of planktonic mayhem. I am back at my desk, looking with idle interest at the polar margins of our latest weather maps, where things appear much different from what they are here at our location. Far to the south, unbroken by continents, is the Southern Ocean, the subantarctic expanse of frigid westerly winds so famous in maritime lore. Here, named for the latitudes they occupy, are the Roaring Forties, the Furious Fifties, and the Screaming Sixties—still the fastest way to sail around the world, if all that concerns you is getting back to where you started from. In the northern hemisphere, more land than water, things are interrupted by the repeated appearance of terrain. Smooth patterns of atmosphere become a packed scrum of air mass boundaries, their symbology evocative of a spider's web or broken car windshield. There is mayhem from Siberia to San Francisco. Below this line the warm stable mesa of the North Pacific High is jousting for position with raucous subpolar air, forming a trail of eddies that are the storms and calms of the temperate regions. What textbooks show as a neat band of westerly wind is in fact chaos—a tide rip of endless turbulence along a line between much different air currents.

We are lucky to be where we are today. Like my shipmates, I have found winter work in the tropics, well aware of what is happening elsewhere as I walk around in sandals. On our last day in Puerto Vallarta my brother called me from Seattle, sheltered in a coffee shop as gales drove sheets of cold rain across Puget Sound and up the western face of the Cascades. He is a chef, raising a family and supporting whatever hope our parents had of sending their children into conventional careers.

I sought shade under the awning of a juice bar as we talked, watching a tourist couple in bathing suits ride past on a tandem bike.

"How's your weather?" he wanted to know.

"Don't ask," I told him.

In Maine, at about the same time, my father shoveled snow off his driveway in a clear blue gale, struggling not to get blown out into the road. The cold front connected to this same system would soon overtake our sister ship in the Bahamas, fifteen hundred miles to the south. Deep in a muggy subtropical night, they were caught by a towering wall of black clouds from the west, moving at 35 knots and stretched from one horizon to another. A torrential line of rain and an hour afterwards a northwest gale sweeping them offshore, the air suddenly dry and cold enough for down jackets.

Here in the trade winds we sail on, warm and indifferent to such distant tribulations.

⁓

CYCLONES are atmospheric eddies that surround an area of low pressure at the surface. They are sometimes called *convergences*, since they all—in some fashion—involve an inward flow, toward a place where air is rising. Nature abhors a vacuum. The converging flow is steered by the Coriolis effect, so that in each hemisphere cyclones spin as would Earth if viewed from the pole: counterclockwise in the north, and clockwise in the south. The term *cyclone* itself derives from the Greek word *kyklōma*—meaning wheel, or coiled serpent—and seems first to have been applied in a meteorological context by a British East India captain named Henry Piddington. Piddington—like Beaufort and the other early giants of marine weather—sought to replace the fractured array of local descriptors with a standard terminology that could be universally applied. The wind at hand might be a *nor'easter, Sumatran,* or *willy-willy,* but all were in a broad sense attributable to the presence of a large rotating low-pressure feature in the atmosphere.

The nomenclature of cyclones can still be confusing, as it is extensive and often loosely handled. An assortment of near syn-onyms is used: *Low-pressure system, depression, disturbance,* and *trough* are all terms for an area where lowered pressure at the surface is drawing air in and sending it aloft. The lifted air expands, cools, and releases moisture through condensation. Clouds form. Rain and snow fall to the ground. These are so-called *synoptic* features, a de-scriptor that indicates their scale and duration. Synoptic events are measured in hundreds of miles and take days to run their course. Embedded in synoptic events are faster-moving subparts; behind them are slower patterns with longer periods of development.

Cyclones sort into two primary families according to their ori-gin. In the middle latitudes, the density difference between warm and cold air creates unstable boundaries that lead to the develop-ment of *extratropical* or *wave* cyclones. This is your garden-variety winter storm—if you live in North America, Europe, or Chile, it is these features that make most of your weather. Closer to the equator, *tropical cyclones* develop in air that is uniformly warm and humid, stirred by the hastened uptake of oceanic moisture at the hottest times of the year.

Most extratropical cyclones get started along the *polar front*, the more or less permanent boundary between the polar and tem-perate zones of each hemisphere. There is a rapid change in atmo-spheric density at this point. Recall that warmer air is lighter than cold, and frequently holds more water vapor—which serves to lower its density even further. The *jet stream*, a high wandering sluiceway of westerly wind, follows the polar front at an altitude of about ten kilometers. Despite its name, the front is usually a long way from the pole. It may reach Florida in the winter and disappear into Arctic Canada by July. Its antipodal counterpart brings storms to New Zealand in June. It can be as straight as a street or as wobbly as a goat path. It may split into branches, or meander back onto itself to leave cut-off lakes of cold air in its wake. Looking down on the front from one pole or the other is most revealing of its character: Seen from here, it wobbles its way around the globe in a jiggly circle, like sauce poured over ice cream. A pattern of peaks and troughs is evident, long sinusoidal curves called *Rossby waves,* which themselves progress slowly from west to east. Turbulence along this boundary induces mixing, and with that, you get weather. A lobe of dense polar air sinks toward the equator, warm air floats above it, and more air rushes in at the surface. Clouds condense as the rising air cools. The Coriolis effect sets the whole business to rotating, and a cyclone is born.

Maritime air is buffered against rapid heating and cooling by its nearness to the ocean, and laden with moisture from evaporated seawater. The most explosive extratropical cyclones are spawned in fall and winter at the eastern edges of continents, where cold dry air meets milder wet air head-on. Frigid winds blow out of Siberia and over the Pacific. The lighter ocean air rises above them like foam on a coffee drink, and more is sucked in to take its place. The rising air is subjected to rapid cooling as condensation converts water vapor to rain and snow. Hokkaido, the northern island of Japan, is the snowiest place on Earth. Upstate New York is not far behind, where winter winds blow in off the Canadian Shield and turn Lake Erie into a giant snow machine. "It is just a little lake-effect snow," they said at a hotel desk near Buffalo, as outside in the parking lot my car disappeared from view.

The polar front makes its way into lower latitudes as fall advances. Gusts of arctic air meet warm Gulf currents off North Carolina. Cyclones develop here, deepening rapidly as they sail up the East Coast. Within a day or two, howling northeast winds lash Boston with rain and snow as the systems move gradually offshore. Momentarily forgotten, they roll across the breadth of the Atlantic, inhaling moisture all the way. By the time they next reach land, they can be hurricane strength, a thousand miles across. The Irish port of Cork is the westernmost harbor in Europe, and the pilot boats here have been photographed diving fully underneath waves on their way to meet incoming ships. At the nearby Kinsale offshore energy platforms, the largest seas recorded each winter are over twenty meters high. These platforms are due for decommissioning soon, but the ocean will no doubt stay in business. I sailed from Cork one July on a voyage to Spain. Even in summer a trip down the verdant Lee estuary and into the sudden vastness of the Atlantic is an arresting transition. It was a peaceful day in the approaches, but I couldn't help noticing how deliberate the pilot became as he prepared to step down the ladder and take his ride ashore, the fields of Éire close in the background and under his feet an omnipresent swell.

⌒〜◞

IKE other fluids, air prefers to stratify rather than mix. Warm air rises above cold. Dry air sinks below moist. Air masses can sit side by side at times, though soon enough some perturbation is likely to upset this equilibrium, and the stirring begins. Extratropical cyclones are characterized by *fronts*, the distinct boundaries between their embedded air masses. Drawn on a weather map, an extratropical cyclone resembles a wave, with its crest centered around an area of low pressure at the surface. The leading edge of the wave is the *warm front*, a wet wedge of warm air that is converging on the center of the low and getting forced aloft by the colder air to either side. Warm fronts are drawn as a solid line with a series of semicircular bumps along the forward edge. The trailing part of the wave is the *cold front*, an advancing wall of dense dry air shown on the page by a distinct line of saw-toothed figures along its boundary. The cold front moves faster than the warm front, and over a span of days the gap between them will close up like a zipper. Once the zipper closes completely, the mixing cycle is complete, and the system dissipates.

The seminal explanation for extratropical cyclones was put forth by two Norwegian meteorologists early in the twentieth century. The Norwegian model was the first to apply the concept of fronts—a term borrowed from the war that had just ended in Europe, and which remains a useful descriptor despite its adoption in the age of biplanes. Jacob Bjerknes and Halvor Solberg were young researchers at the University of Bergen, a location that provided them ample opportunity to experience the foul weather of the North Atlantic firsthand. Their now-renowned paper sums their conclusions up like this:

The cyclone consists of two essentially different air masses, the one of cold and the other of warm origin. They are separated by a fairly distinct boundary surface which runs through the center of the cyclone. In the case of the eastward-moving depressions of the northern hemisphere, the warm air is conveyed by a southwesterly or westerly current on the southern side of the depression. At the front of this current, the warm air ascends the wedge of cold air, and gives rise to the formation of precipitation (warm front rain). The warm current is simultaneously attacked on its flank by the cold air-masses from the rear of the depression. Thereby part of the warm air is lifted, and precipitation is formed (cold front rain).

The Norwegian explanation was robust enough to endure without major adjustment until the addition of something called the Shapiro-Keyser cyclone model late in the twentieth century.

The most notable refinement of Shapiro and Keyser was a clearer vision of what happens in certain powerful marine cyclones, where the cold and warm fronts split apart to allow a slug of warm air into the core of the storm. When seen by satellite, the signature of this process is a great backward-bent comma of cloud, likened to the claw of a hammer or, more dramatically, the tail of a scorpion. As my colleague Captain Jay Amster found in his experience off New Zealand, this can lead to brutal conditions at the surface, as the storm's inhaled heat surges aloft and cold air crashes back to earth at stratospheric velocities.

Observers in the path of a wave cyclone will notice a pattern, starting with the presence of high, wispy cirrus clouds on an otherwise fair day. This classic sign of foul weather is produced by moisture that's been lifted from the center of the system and thrown ahead on the strong westerlies of the jet stream. A day or two after the cirrus clouds arrive it becomes overcast, as a gradually lowering layer of stratus cloud heralds the arrival of the warm front. This thin wedge of warm air may be hundreds of miles across, and the clouds get lower and denser as it advances. Temperature and humidity will increase. It may start to rain or snow. As the system draws closer, the barometer will begin to fall. Wind direction can be used to find the location of the cyclone center using *Buys Ballot's law*, which says that if you face the wind in the northern hemisphere, the center of a cyclone will be to the right and just behind you, at about four o'clock. If the center is west of you, the wind will be southerly, as warm moist air from lower latitudes is pulled in by the convergence.

The cold front is a wave of cold dry air, rushing in to take the place of warm air that's been forced aloft in a wave cyclone. It's an abrupt boundary, more like a wall than a wedge, often defined by tall columns of cumulonimbus clouds and a sharp wind shift toward the polar quadrant. After the cold front passes, barometers typically will begin to rise and the air dries out. It may get sharply colder. Sometimes the cold front is distinct as a windshield wiper, a rapidly moving band that clears the sky in one fast sweep. Other times, if the trailing boundary is more ragged, the passage may involve patchier transitions of cloud and precipitation. My introductory thrashing on the way to Greenland was brought by a classic wave cyclone, formed along a seasonally displaced polar front. We were in effect visiting America's winter weather at its summer home, a broad alley of storms pushed north by the warm air of July. Gales are hatched here when puffs of moist southern air drift off the tail of the Gulf Stream and meet a cold wall of arctic air. Drawn in by the storm's warm sector, we were halted at the blunt anvil of its cold front, the north wind building an abrupt sea and eventually delivering the clear skies of the subsequent dawn. I was twenty-eight at the time, briefly past the threshold of my captain's career and sustained more by adrenaline than by experience. I'd been given the opportunity to sail a famous vessel to an exotic place and I'd taken it, just barely aware of all it would involve. This is how adventures begin, some will offer. Adventures are for the unprepared, others might respond. Both would be correct. On my last shopping trip before sailing, I spent $200 on socks and underwear, and everything I had left on Snickers bars.

⁓

THE tropics lack the sharp contrasts in temperature that drive weather in places like Duluth or Tokyo. Here instead the near-vertical sunshine heats Earth continuously, forming a belt of warm wet air where trade winds push their way steadily in toward permanent zones of convergence near the equator. Tropical cyclones develop in this environment when elevated sea temperatures trigger pockets of locally low pressure in the atmosphere. The mature tropical cyclone is the hurricane, the final peak of a pyramid that may or may not be built to completion on any specific occasion. *Hurricane* is itself a term specific to the Atlantic and eastern Pacific, traceable to an indigenous Taino word from the Caribbean basin. By legend, Hurakán was an angry wind goddess, shown in local iconography as a sort of wide-eyed dervish with long whirling arms. She looks more than a little like the satellite image of a tropical storm, implying that the Taino in their pre-European world already had a good grasp of the local meteorology. Analogous storms in the western Pacific may be called cyclones, typhoons (from the Chinese *tai phung*), or bagyo, a Tagalog word from the oft-impacted Philippines.

Tropical storms often originate as disturbances in the trade winds, where an area of localized convection expands and begins to rotate. Air is drawn in toward the center, lifted, and carried away by winds aloft. Condensation occurs in dense concentric bands of cumulonimbus clouds, and the storm takes on the appearance of a spiral galaxy made of thunderstorms. A steep and sudden pressure difference develops between the edges of the feature and its center, shown on weather maps by a bull's-eye of closely packed isobars. Amid the otherwise gentle grain of wavy lines, a hurricane stands out like a knot in a pine board. Tropical maritime air carries huge amounts of stored energy in the form of latent heat, entrained in its vast cargo of evaporated moisture. When that water vapor condenses into clouds, heat is released to the air column, strengthening the convective cycle. This is what scientists call a *positive feedback mechanism*—where the outcome of a process encourages the process itself to continue. This release of latent heat is a primary catalyst in weather systems of all kinds. It's what makes clouds grow tall. The tropics, however, are unique in their bottomless store of warm moist air. Given the right circumstances, a benign spot of localized convection can harvest enough energy from water vapor to expand into a tropical cyclone.

In the North Atlantic, aspiring storms are visible in the form of *tropical waves*, traveling wiggles in the trade winds that pass across the ocean at the rate of about fifty per year. Most come to nothing, though about a third will grow eventually into cyclones. Viewed from aloft, they appear as brief undulations in the easterly flow, small troughs of equatorial low pressure that push north by a few degrees of latitude. Some are durable enough to blunder across Central America and into the eastern Pacific. If you've sailed in the tropics, you've probably been in one—a persistent band of patchy rain and gusty wind, accompanied by a dip in the barometer and a series of shifts in the wind direction. In the western Pacific, cyclones form in the South Pacific Convergence Zone, a vast migratory gutter of disturbed air at the confluence of the trade winds and something called the Kermadec High. Here clusters of local thunderstorms can become dense enough to develop into organized disturbances, some of which gather sufficient energy to become cyclones in their own right. I watched this process closely for the first time in 2015, when a knot of rain near the Solomon Islands coalesced into Cyclone Pam just as we were preparing to leave the port of Dunedin on New Zealand's South Island. A storm of lethal intensity, Pam mostly spared the Kiwis but did horrible damage to the island nation of Vanuatu, where at least sixty people died. In the news were images of a once-lush jungle landscape, transformed by wind into a clear-cut.

Science is still without all the tools to say exactly when a tropical cyclone will develop, but a sort of checklist exists: The sea surface temperature must be at least 26 degrees Celsius, a limit that confines the birthing of storms to a relatively narrow band of ocean. One must be far enough from the equator—generally at least 5 degrees of latitude, or three hundred nautical miles—to be influenced by the Coriolis effect. And in a way that is hard to see unless you are a scientist looking at balloon data, there must be the right regime of winds aloft: brisk enough to carry lifted air away from the center of the storm, but not so brisk as to disperse the system itself. The Hurricane Discussion—perhaps the least charismatic product among the National Hurricane Center's many offerings—is a nerdy block of plain text written by meteorologists as running commentary for each storm. It is dominated by talk of things like "shear," "exhaust winds," and "vertical symmetry"—all high-altitude phenomena that are the backstage actors in surface weather events.

Tropical cyclones occur in all of the ocean basins, though they are more common in some than in others. They are most likely from late summer to early fall, when seasonal heating pushes sea temperatures to their annual peak. Moist air, warm ocean, and favorable winds aloft will allow a storm to build through progressive stages until it reaches hurricane status—with winds above 64 knots and a characteristic *eye*, the nucleus of calm air that is formed by the closed ring of intense convection just outside. At sea, the eye is a spot notorious for chaotic conditions as waves are driven in from every direction by the surrounding circulation. The most intense winds occur in the *eyewall*, a thick ring of cumulonimbus cloud just outside the eye itself. From there things moderate gradually with distance from the center. Compared to a nontropical cyclone, the radius of full-strength winds in a hurricane can be fairly compact, an average of fifty miles. The area of dangerous winds, normally considered as those above gale strength (35 knots), might extend to two hundred miles.

The Saffir-Simpson Hurricane Scale starts at category 1 and goes as high as 5. Category 5 storms have sustained winds of at least 135 knots, and a central pressure below 920 millibars. Storms this powerful are relatively rare—according to the National Hurricane Center, there have been only about forty in the Atlantic Ocean since record keeping began in 1851. Hurricane Patricia, formed in the eastern Pacific in 2015, is the current record holder for intensity, with peak sustained winds of 187 knots. Wind speed, however, is only one measure of a storm's destructive potential. Hurricane Mitch dropped seventy inches of rain on Honduras in 1998 and is connected with as many as ten thousand deaths related to flooding. In August of 2011 the sodden remnants of Hurricane Irene turned the mountain town of Rochester, Vermont, into an island. The city of New Orleans was famously devastated by Hurricane Katrina in 2005 through a combination of torrential rain and *storm surge*—a wave of windblown seawater pushed along by the advancing system that increased local sea levels by as much as twenty feet.

If a storm surge arrives together with the high tide, it can be especially dangerous. Someone I know was the waterfront superintendent at Manhattan's South Street Seaport Museum in the fall of 2012 when Hurricane Sandy arrived. He and his crew had just finished rigging storm moorings when the tide came in. The water rose six feet in an hour. It pushed some pontoon docks up into the parking lot, crested a bulkhead, and kept going. South Street became a river. When the crew saw green water begin to pour down the stairways leading to subway stations, they knew that much more serious problems were on the way.

Predicting where hurricanes will go and what they will do remains an inexact science. As with other storms, hurricane forecasters do their work with a combination of observation, computer modeling, and intuition. The tenacious unpredictability of hurricanes is best encapsulated in something known as the mariner's 1-2-3 rule, an old empirical tool still in wide use by sailors. The 1-2-3 rule works like this: When plotting a series of forecast positions for a storm, a mariner first draws a circle around each one to represent the expected radius of dangerous winds. Then, to account for uncertainty, they add an extra sixty nautical miles to the storm radius for each day into the future that the forecast projects. The resulting field of expanding circles forms a graphical boundary of possible positions for the storm over the next seventy-two hours, all places that you don't want to be. Expanding this process into a fourth day might produce an area large enough to fill half an ocean basin, another way of saying that any long-range forecast of a hurricane's movement should be treated with extreme caution.

Very generally, tropical cyclones take parabolic paths—moving slowly west while in the tropics, and then re-curving to the east as they travel into higher latitudes, steered by the prevailing westerlies and a stronger Coriolis effect. Often they will skirt the large resident high-pressure systems in the center of the main ocean basins, following the warm currents at their western margins. The Gulf Stream carries tropical water and warm humid air up the US East Coast in summer, a highway for hurricanes. The Kuroshio Current does something similar in the western Pacific, where hurricanes are called typhoons. Storms in the North Atlantic tend to start on a westerly course, aimed first toward the Caribbean and then for mainland North America. Eastern Pacific storms— which get less publicity than their Atlantic counterparts—form west of Central America and may curve toward Hawai'i or take sharp right-hand hooks back into Mexico, where the paths of old hurricanes are visible as green streaks of vegetation in the desert.

The first complete plot of an Atlantic hurricane track is credited to an American steamboat owner named William Redfield, who collected myriad local observations of a storm that had punished the eastern US in September of 1821. Captivated by the realization that winds from opposite directions had been recorded simultaneously in the neighboring states of Connecticut and Massachusetts, Redfield used his data to uncover both the path and the profile of the storm—correctly characterizing it as a "traveling whirlwind" that over several days had made its way from the West Indies to Nova Scotia, trailing destruction in its wake. His conclusions led him into unexpected conflict with a scientist from Philadelphia named James Espy, who was building his own theories on the workings of convection and evaporation. Espy had a different model for storms, built around the idea of wind rushing straight in toward a rising central column of warm air. He and Redfield spent the next thirty years in argument over who was wrong, both sadly passing just as work by a mathematician named William Ferrel showed that each had been holding separate parts of the same correct explanation: Hurricanes—and cyclones in general—are in fact the joint product of convergence and rotation, imparted in turn by the relative buoyancy of warm air and the torque of the Coriolis effect.

A HURRICANE in the northern hemisphere will as a rule be more intense on the side that is to the right of its track, where the momentum of its forward velocity is added to the wind strength. Textbooks speak of the "dangerous" and "navigable" semicircles (as if one would ever think any part of a hurricane as nondangerous), but mariners confronting these storms know that there are bad and worse places to be. A second, less obvious hazard of the dangerous semicircle is that the circulation therein carries vessels back into the path of the storm, rather than away from it.

Captain Phil Sacks is an old shipmate from the sailing world who had the unwelcome opportunity to think about all of this some time ago on a voyage from Cape Cod to the Virgin Islands. On the afternoon of October 22, 1992, he and his ship—the schooner *Westward*—were northeast of Bermuda, slowly returning to their routine after being bounced around by a gale in the days preceding. October is still hurricane season in the North Atlantic, and *Westward* was keeping well east in order to stay clear of the most common late-year storm tracks. Despite this precaution Captain Sacks found himself facing the sudden news of a hurricane named Frances, forming less than three hundred miles south of his position. This was a long way north for a fresh tropical storm to be developing, but it's not unheard of. A large, energy-laden mass of air had pushed up out of the tropics and over the warm Sargasso Sea. A relaxation of the winds aloft allowed it to converge into a rotating column of convection, and a cyclone was born. This might not happen in June, but after a long hot summer the sea surface temperatures were at their peak and all necessary ingredients thus at hand.

The captain was surely uninterested in these nuances at that moment. He needed to get his ship out of the way. Frances was forecast to head due north, and at first Phil's plan seemed like a sound one: Make tracks to the southeast, working diagonally away from the storm's projected track while its center passed astern. This would take *Westward* through what was nominally the dangerous side of the system, but Phil felt like he could achieve adequate margins as long as the ship made good time and the storm stuck to its forecast. Unfortunately, neither of these contingencies came to pass. With Frances's center located southwest of the ship's position, the winds were from the southeast, and progress became steadily more difficult as the day went on. Turning downwind would make sailing much easier but bring the ship directly back into the storm's path. And in the wee hours of October 23, the hurricane itself altered course, beginning a curve toward the northeast. The way things were going, *Westward* would now pass within just fifty miles of the eye, provided she could even maintain her current speed in the building seas.

Captain Sacks thus found himself facing all the threats of the dangerous semicircle. With his ship to the right of the storm track, seeking escape in the direction he was headed meant fighting for progress upwind, all the time slipping back into the hurricane's path. And Frances was in statistical terms very likely to continue turning to the east. It was a warm black night. The winds were still below 35 knots, but the barometer was falling hard and the sky filled with lightning. Racing across the path of an oncoming hurricane is a frightening prospect, but that is exactly what the captain chose to do next. He went on deck and gave his decision to the watch: *Westward* would do a one-eighty and turn west under reduced sail and full engine power. This would take her directly across the hurricane track, but at maximum speed and with the wind astern. If the forecast held, she would clear the eye by a hundred miles, and be in the safer left-hand semicircle by morning. I can picture the ship's engineer standing by in this moment, holding on with one hand like a rodeo cowboy and striving to adjust the throttles smoothly as the vessel pitched through its forty-foot arc. Phil not far away, pulling hard on a cigarette kept alight by sheer will in the driving rain.

"Warp speed, Mr. Sulu," someone must surely have said.

It was a wild ride, but ultimately a successful one. I heard fragments of the story from friends in the months that followed, and a few years later a complete version emerged in a textbook written by Captain Andy Chase, a professor at Maine Maritime Academy. Here is part of Phil's account, begun not long after his decision to change course:

Oct 23—I woke and noticed it was not quite dawn. I suited up in rain gear to check the situation on deck. The winds had built to force 8 or 9. What really caught my attention was the wind direction. At midnight the watch had recorded wind from the southeast. Shortly after we began running west, at 0200, the wind had backed to the east. It was now 0430 and I felt the wind gusting from the northeast. A backing wind was exactly what we expected to feel if we were in fact crossing the track into the navigable half of the storm. The barometer was now falling nearly 2 millibars per hour, but the backing wind lifted my spirits. I nervously, but optimistically, headed below to catch the 0600 weather broadcast.

The report began . . . I was again struck by how quickly this storm had developed. We had been involved with Frances for only 16 hours. The eye's position was given first: 30.0° N x 60.0° W at 0500—now tracking 030° true at 10 knots. We had made it! We were in the navigable semicircle. I knew that we were far from in the clear—that the storm could at any time change its track again. But Frances had taken the expected track. We would now be able to distance ourselves by running with the wind on our starboard quarter. I felt a tremendous weight lift from my shoulders. The wind was still building, but we faced it with more confidence. The sun was rising to the east, an incredible relief. Through the morning the wind and seas continued to build. By 0800 we were logging winds of force 10, perhaps 11. At that point it is so difficult to know. The largest seas were 20 feet. They looked somehow different from most other storm seas I have encountered. They appeared glassier, more smooth-surfaced. Perhaps this was due to the tops being blown off by the wind. The ship continued to ride well as we ran before the storm at better than 8 knots. The barometer had reached its nadir. We were now distancing ourselves from the hurricane.

In retrospect, Captain Chase admits some ambivalence about this story as a teaching example, featuring as it does the risky stratagem of crossing a hurricane's path to gain safety in its less dangerous semicircle. Such a move is surely written somewhere on a list of things that a sailor must simply never do. To me, in the nuanced world of actual seafaring, it is an example of what may turn out to be your best option amid a set of bad alternatives. Captain Sacks saw a hurricane form suddenly, south of his ship and headed his way—and after determining that getting east of it would be impossible, he made a prompt, if bold, decision to race back across its path. In doing so he bet correctly that the storm would be turning east as he went west. This last was a critical contingency. If Frances had instead stalled or turned the other way, the outcome might have been much different.

More tragic is the story of the tall ship *Bounty*, whose master, Robin Walbridge, took the inexplicable decision to leave New London, Connecticut, in an attempt to outrun Hurricane Sandy. It was October 26, 2012. Sandy was pelting the Bahamas, forecast to track north in the next seventy-two hours on a path more or less parallel to the US coast. As the day drew on toward evening and other ships made preparations to ride the storm out in harbor, Captain Walbridge told his startled crew that he planned instead to sail immediately and race offshore of the coming hurricane, sprinting east to gain sea room while the system passed behind them.

Regrettably, once at sea, Walbridge quickly found himself facing a plight much like what Phil Sacks had experienced in his attempt to sail east of Hurricane Frances. Even as *Bounty* fought to gain ground, headwinds in the cyclone's upper right quadrant pushed the ship steadily back into the storm's path, making it all but impossible to skirt the system as planned. From here, the captain decided instead to sail back inshore, recrossing the storm track and turning to run south with a tail wind in Sandy's left semicircle. This choice had problems of its own. Sandy was weak as hurricanes go but covered a very large area, growing larger still as it moved out of the tropics and began to ingest colder air. In defiance of some early forecasts, the system made a turn back toward land on October 28, drastically limiting the room available for any vessel trying to pass along the coast. The winds built to 70 knots and seas grew to thirty feet, breaking chaotically amid the countercurrents of the Gulf Stream. *Bounty*, an old wooden ship in questionable repair and leaking badly, suffered a terrible daylong spiral of crew injuries and equipment failures before foundering off Cape Hatteras early on the morning of October 29. Her people, thrown into the water as their vessel capsized, managed to set off a distress beacon and were miraculously located by US Coast Guard aircraft. All but two were saved: A deckhand named Claudine Christian was recovered by rescue swimmers but could not be revived. Captain Walbridge was not found, and his decision to leave the safety of port never fully explained.

At a conference that winter a tribute was held for the lost sailors and their ship. Several of the surviving crew were in attendance. None had any real idea what their captain had been thinking, other than to say that they'd seen him accomplish remarkable things before and perhaps were inspired by his boldness to follow as they did. Also in the building were members of the Coast Guard air group that had performed the rescue, invited for an award ceremony in recognition of their service. At a party afterward I met a pilot who had flown his C-130 to the scene, racing under the cloud deck of a hurricane and squinting out his cockpit windows for signs of life. It was just before dawn. The aircrew had spotted the doomed ship and then a lone strobe in the water, floating separately from the life rafts—a light that on later inspection would turn out to be *Bounty*'s first mate, clinging by himself to a bit of buoyant debris.

Sipping a drink that someone had bought him, the pilot explained to me the job that had followed, which was to bank his plane immediately and get back around to the scene. That is to say, turn a forty-ton aircraft in a circle tight enough to maintain contact with the rescue target, all without getting lost in the clouds or diving into the ocean. He spoke in the calm voice of aviation—a bland twang, evocative of places far from salt water. His plane had circled the scene while the rescue took place, a flying marker for the two helicopters that hoisted people one by one from the maelstrom in mesh baskets, hung from a thin strand of wire.

Outside a blizzard swirled around the windows as light drained from the sky. We were in Erie, Pennsylvania—a place not known as a mecca in maritime lore. The pilot emptied his glass and glanced around for his mates. He looked very young, somewhat bemused at his sudden celebrity. There was little I could think to say in response to his story. On some days without warning you meet the people you most aspire to resemble, and in following can only strive after their example.

Tropical storms will persist as long as they have favorable winds aloft and a good supply of warm wet air. They usually dissolve after drifting ashore, though it may take days for their energy to dissipate, and they remain dangerous well after landfall. Storms can reignite if they move back out over warm water, as often occurs when they strike the Florida Gulf Coast and then reemerge over the Atlantic. In the final chapters of its life, a hurricane may transition to a *post-tropical* status, mixing with colder air and developing frontal boundaries like a wave cyclone. Such systems often have a vastly expanded footprint, with persistent winds above gale strength throughout. It's not unheard of for a tropical storm to be swallowed whole by another nontropical cyclone. This abrupt injection of heat and moisture can yield dramatic results, like Popeye with a can of spinach. It's what happened with the "Perfect Storm" of 1991, which was created by the explosive union of a late-season hurricane and a fall nor'easter off New England. Hurricane Sandy went through a similar evolution as it merged with an extratropical system just before reaching the New Jersey coast, putting the Ocean Prediction Center in the rare position of ordering blizzard warnings for a hurricane. I recall a remarkable satellite image of Sandy's cloud disc headed right into the jaws of the northern cyclone, looking most of all like a screen shot from an old Pac-Man video game.

It can be difficult during these transitions for forecasters to sort a storm clearly into one category or the other. They may share a blend of tropical and extratropical characteristics—or the issue may be one of management. The National Hurricane Center in Miami is responsible for tropical storms in the continental US, while warnings for wave cyclones are normally handled by various branch offices of the National Weather Service. When a tropical storm dissipates or becomes extratropical, its ownership may become unclear. The preference of the National Hurricane Center at the time of Sandy was to hand such storms off to their non-tropical colleagues like air traffic on its way from Miami to JFK. This created a problem for emergency managers when the storm started this transition mere hours before its projected landfall on the New Jersey shore. Was it still a hurricane? Did it make sense to hand over forecasts and warnings in the midst of a storm's arrival at North America's densest population center? What would be least confusing to emergency managers?

Negotiations ensued. After much discussion on scene, Sandy's career as a hurricane ended administratively, and all warnings north of a chosen point—Duck, North Carolina—became non-tropical. Meteorological fine points stepped aside in favor of efficient disaster management. Or so some thought. Others felt that without the gravity of an official hurricane warning, preparations for the storm were thrown into some uncertainty. The aftermath of Sandy has led to a paradigm shift, whereby specific offices assume a cradle-to-grave ownership of such systems for as long as warnings are merited. If a storm is born in your house, so to speak, it stays your problem.

Tropical cyclones are given their names according to a range of customs around the world. The practice at NOAA until 2021 was to alternate between boys' and girls' names—Bonnie, Carlos, Deborah, etc.—and move on to Greek characters when the annual supply of storms outran the number of easily matched letters in the Roman alphabet. Such was the case in 2005, a record year of twenty-eight cyclones in which category 5 Hurricane Wilma was succeeded by storms Alpha, Beta, Gamma, Delta, Epsilon, and Zeta. This was a season of the sort we all thought we'd never see again, until it was outdone utterly in 2020 with a count of thirty-one—culminating with a category 5 assault on Nicaragua by Hurricane Iota. I had never even heard of the letter iota before its namesake popped up on the weather maps. Others might have admitted to similar struggles with ancient alphabets, because the National Hurricane Center announced not long afterwards that it would give full human names to all future storms, regardless of the overall count.

At any rate, the list of cyclonic monikers is prepared six years in advance and recycled thereafter. Historically significant storms have their names retired, and for this reason we won't see Mitch, Sandy, or Katrina again. The letters *Q*, *U*, *X*, *Y*, and *Z*—perhaps viewed as too tricky to find names for—are also left off the list, so we'll never see the likes of Quinn, Ursula, Xena, Xavier, Yolanda, Zelda, or Zebulon either. A pity. Names aside, storms are remembered mostly by their impact on populations, and without the prospect of significant death or destruction, you are unlikely to hear much about them on the Weather Channel. Does anyone in New York City remember Hurricane Frances? Consider the

Wikipedia entry for the improbable tropical storm Zeta, which formed in December of 2005 and actually lasted into the new year.

"As Zeta never approached land," goes the text, "there was no impact from the storm other than minor shipping problems."

As a mariner by trade, I wondered if the shipping in question felt Zeta to be just a minor problem at the time. I'm well-conditioned to long hours spent poring over offshore forecast materials—maps and discussions concerning parts of the planet where only sailors and birds are likely to be found. But to most who watch the weather, the ocean is a place where gales can safely go without disrupting their lives or threatening their property. In his book *Looking for a Ship,* the author John McPhee enshrines a kind of tandem cliché, recounted in this case by a seaman on the merchant ship *Stella Lykes:* A sailor, home from the sea, is stretched comfortably in his favorite chair and watching the news—whereon a forecaster describes the aftermath of a destructive coastal storm in New England. There have been great tribulations in the cities and towns, damage to property and perhaps even fatalities, but now, says the weatherman, the danger has passed.

"The storm," he assures, "has gone safely out to sea."

5

GOD'S ROOF

Some indigenous methods of voyaging reverse the Western concept of motion, using instead a system in which the navigator departs in their canoe, watching land disappear astern until eventually—over a span of time that might involve hours, days, or weeks—another island appears ahead, pulling slowly into view to replace what has been left. Through the interim it is the sailor who inhabits the center of a fixed frame, one where the routines of the day—the ship's chores, navigational tasks, and social interactions—form a fulcrum around which the rest of the world revolves. This is to me an affecting and not entirely unfamiliar notion. Asked to choose, I will generally submit a preference for long voyages over short ones, passages where between the heavy lift of departure and the anxious run-up to arrival is an interval during which the ship and its routine can simply be where you are. I tell this to the trainees who clamor to know of our destinations, unaware that the most compelling thread of their experience will be their occupancy of the ship's world—momentarily but completely divorced from any other reality, real or virtual.

I haven't spent enough time in large commercial vessels to say if this sentiment applies across the industry. Probably not, but among sailors it surfaces frequently. My wife is an environmental engineer but was once also a mariner. Between us we have circumnavigated the world in mirrored halves. She has climbed the pyramids, swum with whale sharks, and slept on piled grain sacks under armed guard in the African port of Djibouti. For all of this, most durable in her memory are recollections of endless tropical nights at sea, crossing the Indian Ocean in a leaky Dutch ketch—pumping every hour and listening to selections from a short list of endlessly looping conversations on punk bands, Formula One racing, and the ever-unresolved question of whether Harrison Ford's character in *Blade Runner* was in fact a replicant. The passage from Australia to Suez took three months, much of the last spent sweeping thin shoals of red dust off the deck each morning, gritty bits of an invisible desert.

Dodge Morgan, a fellow New Englander and late acquaintance, made his fortune in electronics but is best known for a record-setting circumnavigation that he completed in 1986 aboard the yacht *American Promise,* built specially for the purpose. It took him 150 days, during which he celebrated his fifty-fourth birthday. A popular speaker afterwards, he would step to the podium at this or that club dinner and wait for things to get quiet.

"Trying to get somewhere by sailing," I once heard him begin, "is fucking stupid."

His point being, I think, that anyone setting off on such a course would be best served to focus on the process, not the outcome. It evokes the story of Bernard Moitessier, a Frenchman who in 1968 was poised to win the first ever round-the-world solo sailing race when he decided that he would instead just keep going. He stopped eventually in Tahiti, another half trip around the world.

Occasionally I'll sit to leeward of my own charthouse and look out at nothing, vertiginous moments in which it's possible to imagine that the water is rolling by like a carpet and we are standing still. In these instants it is sometimes possible to gather the whole ship in my mind's eye, all the moving parts and human routines but also our long line of travel, the interval during which the vessel is its own world entirely—a capsule in free fall, unaffected by the gravity of either origin or destination.

The length of this voyage is great enough to make our goal feel close when it is still three hundred miles away, two full days of sailing if we hurry. Nuku Hiva sits midmost amid the Marquesas archipelago, a cluster of dots populating the chart like sand grains near the 139th meridian, six hundred miles south of the equator. Their storied appellations are said to name the subparts of a house under construction by the Polynesian god Atea: Hiva Oa, Nuku Hiva, Ua Huka. God's ridgepole, God's roof, God's woodpile. First on the list of today's novelties is that we are near enough to be navigating on a chart at all. Offshore navigation is traditionally performed on a plotting sheet, simply a blank page imprinted with the grid of latitude and longitude in thin green lines. Nothing more is needed, since absent any known obstacles (say, Hawai'i) the abyssal ocean can be presumed deep enough for any vessel to float in.

Assuming that a ship keeps a proper lookout and remembers when it is near the odd island chain, this sort of deep-space reckoning is considered generally sound from a safety standpoint. This said, the oceans are vast—particularly the one we are in at the moment—and even after a millennium of human transits there are still things to give one pause. The chart we have rolled out for our arrival is a British Admiralty product, bearing a catalog number perhaps first assigned by Beaufort himself. Who knows? It is an old edition, rendered in crisp monochromes as if by some mustachioed royal engineer hunched by his inkwell.

Displayed in this spare medium, the place where we are going looks not unlike a hand-drawn map of Treasure Island.

Embedded in chart margins is often a smaller window, a sort of mini-chart showing quilted outlines of the individual surveys included in the overall mosaic. Not all will be of recent origin. "From surveys by NOAA ship *Ferdinand R Hassler*, 2016," one block might say, right adjacent to "Soundings by HMS *Whippet*, 1872." The ocean is a big place. Our penumbra of knowledge grows ever larger but remains incomplete, and charts of lightly traveled areas may still hold terse warnings to the navigator in their margins:

"Suspected Volcanic Activity, 2014."

"Snook's Island is reported to be two miles east of its charted position."

You get the idea. On this day's chart, typical of the vast Pacific, one finds minute dotted circles far from anywhere, tiny labels adjacent thereto:

"Breakers, reported 1912. Existence doubtful."

A breaching whale? The basaltic brow of some unknown volcano? *Vigia*, these spots are called in navigation, from the Spanish: a navigational hazard whose existence or position is uncertain. Caveat emptor.

The Marquesas are dead downwind now of our position. We are closing the distance in a series of long diagonals like an airplane letting down its altitude on approach to a field far below. Our squaresails stretched against the sunlight. For the sailor, ground held to windward is like altitude to a pilot—a priceless reserve to be emptied only with the runway in sight, its numbers readable from the cockpit windows. On this passage we have been generously to windward of our goal for the whole distance, all the while knowing that through inattention it remains still possible to overshoot, to give up precious advantage and be forced to work unnecessarily to regain ground that was once ours.

Such a development would be for us embarrassing, though in truth not disastrous. We can with the aid of our main engine regain considerable lost ground if obliged to do so. In the days of pure sail this would have been a much different story, and the boundaries of modern geopolitics are often laid out around older realities of who was to windward of whom. Consider Barbados, in the next ocean over from ours. The Lesser Antilles of the Caribbean are a long jeweled necklace of volcanoes, an island arc extruded from the tangle of plate geology beneath. Their written history is commentary from a long-running game of colonial ping-pong, holdings traded repeatedly between foreign powers before achieving their final autonomy in the twentieth century. Not so Barbados, which through a tectonic anomaly sits one hundred miles east of the main island chain—meaning that any approaching ship has one chance to find it before being swept hopelessly past to leeward. Barbados became British in 1625 and stayed that way until its independence in 1966, a year after I was born. Britain itself has been spared from a significant seaborne invasion since the Norman conquest of 1066—in large part because any would-be assailants from the continent were as a first step obliged to load their armies onto ships and proceed upwind against the prevailingly horrible weather of the North Atlantic. Faced with such nonstop adversity, the sailors of northern Europe have as a consolation emerged to be the best in the world—an accolade they would share with the Kiwis, another maritime society who have developed their craft amid a steady pummeling from violent and capricious winds.

In our chartroom a modest array of instruments is clustered around the dim glow from the chart table, lights tinted red to preserve night vision and help us feel like we are the navigators of something racier than a mere brigantine plodding across the blank Pacific at 6 knots. A nuclear submarine, perhaps. Above the chart are the silent radios, a GPS, and a pair of radars, which for the moment are empty, even when toggled out to their farthest range. There is nothing to observe on their screens but faint sea clutter—the occasional grainy return from wave tops—and a brief brightening of the heading markers each time the sweeping beam of radio waves comes back up to zero.

There are no other ships here with us today in what sailors call the "lonely longitudes," and the air feels oddly dry, bearing few clouds substantial enough to make rain. We are under the effects of the Humboldt Current—an enduring trail of cold upwelled water from the coast of Peru, swept into the counterclockwise circuit of the South Pacific Gyre. It is an enlarged replica of the South Atlantic, where the Benguela Current carries cold water away from the rich fishing grounds of West Africa and toward Brazil. The result in each case is a corner of the tropics uncharacteristically low in humidity, with much less of the transient squall activity typically connected with low latitudes at sea. Tropical storms are nearly unheard of in these places, and this is why.

❧

T HE Marquesas when we find them have bold green windward shores and long arid rain shadows on their sheltered western coasts. In Nuku Hiva the anchorage at the main settlement of Taiohae is a bowl-shaped indentation in abrupt jagged hills, tall brown hummocks with verdant seams of bush following watercourses up to daylight. Stands of hardwood run the ridgelines in a thin fringe. With the crew I walk the single concrete road, which switches back and forth up the grade from town, past tidy houses with gardens of bougainvillea and mango. Ancient roadside banyan trees are flanked with worn standing stones—some carved in the likeness of Tiki, the first man of Polynesian origin stories.

Up high there's grassland, and broad spare acacia trees. Cows and horses roam free. It's hot until we reach the hillcrest, where a cloudy puff of north wind meets us like a drink of cold beer. It's suddenly greener up here, almost a cloud forest, with tall tree ferns, pines, and epiphytes growing in the crooks of branches. The road divides, marked by neat signs, and tacks back down into valleys that look as impassable as they are breathtaking, with sharp gullies and foliated peaks like spikes of meringue. Down one such valley is Taipivai, the village where Herman Melville spent a month in 1842 before conjuring his debut book, the quasi travelogue *Typee*. The airport, I will learn, is thirty miles from here by road, a two-hour drive. Spread flat, it feels as though the ground between might cover Texas.

Before shore leave can begin, I must first bring our paperwork to the gendarmerie—a perfect French police station parked in a banana grove, a spotless blue Land Rover in the driveway outside. In my backpack are precious passports, a stack of customs forms, and a handheld radio barking random bursts of static. Behind me our inflatable dinghy races away, towing its foamy wake back toward the ship and breakfast. There is bird noise in the trees, and what I realize is a disconcerting feel of solid ground underneath my feet, my inner ear racing to make adjustments.

Nobody is to be seen at the gendarmerie, the morning sun already strong. I find a shaded bench and look back toward the *Robert C. Seamans* riding at anchor. She looks very far away. This is my tendency after long passages—I bring the ship in to a spot that I find close enough for comfort and then have the sense we've somehow anchored ourselves in mid-ocean, an unnecessary waste of time for the boats bringing people ashore. The sight of land is welcome yet somehow unsettling when you haven't seen it for a while. Sailing my home coast, I will pass a stone's throw from the beach and think nothing of it, the smells of bait and diesel exhaust drifting freshly to us from the fish piers. We look a bit like a fishing boat ourselves, I think. After a month at sea the ship's hull is dull with dried salt and faint streaks of rust bleeding from the scuppers. She's carrying a slight list to starboard, left by the constant shuttling of fluids between a dozen different tanks: fresh water from the desalination plant, effluent from sinks and toilets stored for treatment and eventual discharge. Also there is precious diesel fuel, burned at a steady drip by the generators and in great gulps by our main engine when we motor.

Keeping track of all this is a big part of what engineers do at sea, the arrangement of liquids in tanks being connected directly to the stability of a vessel—that is to say, whether or not it will float upright, and under what conditions it will recover from the adverse forces of motion. Full tanks, generally, are better than empty ones. Partially full, or "slack," tanks are considered worst of all, allowing as they do a cascade of their contents toward whichever side of the ship is already leaning downhill. Naval architects call this *free surface* in their calculations. In our ship's papers is a stern missive from the Coast Guard reminding us to maintain the minimum possible number of slack tanks, in general no more than one from each family: diesel, effluent, potable water. For cargo ships the proper management of tanks is critical, as a vessel loaded incorrectly might fracture or capsize from the imbalance. In July of 2006, the car carrier *Cougar Ace*, stuffed with thousands of new vehicles bound for America, mishandled her ballast water and flopped over abruptly at a 45-degree angle—floating abandoned in the Gulf of Alaska for more than a week before being salvaged. Less fortunate was the *Golden Ray*, a ship that suffered a similar casualty while departing from Brunswick, Georgia, in 2019. In a critical error, the *Ray*'s crew had stowed a fleet of heavy SUVs on an upper deck and left several key ballast tanks empty below. Parked on a mudflat and forever finished with her sailing career, she was sold for scrap—sawn into slices like a meatloaf and hauled away by barge in an operation that would take two years to complete.

Our list today is more of an aesthetic issue, I think—perhaps even an enhancement to our rakish silhouette as we float in the outer harbor, yards braced square and flags flying like something from a century ago. I will speak to David and his assistant when I get back and see what they think. When I left, they were busily at work crafting Christmas presents and may have other priorities.

A gendarme in a crisp white shirt appears from somewhere and unlocks the door to his office.

"Bonjour, monsieur."

He is in full uniform, with a cap and sidearm, and could be in Paris, though from his appearance is likely a local. We are near one apex of what is sometimes called the Polynesian Triangle, a broad ethnic polygon stretching roughly between New Zealand, Hawai'i, and Rapa Nui (Easter Island). The Marquesas mark the outer limit of *French* Polynesia, a vast holding that itself runs from here back to Bora Bora, eighty miles west of Tahiti in the Society Islands. Altogether there are five discrete island groups in French Polynesia (not counting a bizarre scrap just west of Mexico called Clipperton Atoll), and their collective footprint gives France the world's largest exclusive economic zone—7.3 million square miles—just topping the US and nearly double that of England.

The fact that these places are both French and Polynesian is to me a case of two equally implausible circumstances: that the French could overtake this utterly un-European terrain three centuries ago and leave a mark so enduring as a Marquesan in a gendarme's uniform, or that the Polynesians found their way here five hundred years before that, sailing open boats across endless water, using skills nearly lost to time. The recent renaissance of traditional navigation has proven the concept and recovered some of the methods likely used, but the fact remains no less awesome to contemplate.

The first Marquesans are thought to have come from Tahiti or thereabouts, eight hundred miles away, on what typically would be a voyage directly upwind. More than likely this means that they understood climate patterns well enough to know that certain years held the rare chance to sail with the wind astern, and in the presence of the right signs set forth. Once at sea they used the stars as a compass and let the rhythm of swells reveal the influence of currents and invisible geography. Sometimes birds were their pilots, until the point where—as with ourselves a millennium hence—their destinations were revealed by tall clouds built from winds lifted over land, the abrupt monoliths of volcanic rock shrouded by billows cast pink in the low light of sunrise.

The deep history of these achievements is being teased out today by a vibrant community of traditional sailing societies, and the sharp pencil of isotope dating. The precolonial population of the Marquesas is figured to have been close to a hundred thousand, scattered among these dozen or so islands where they fished, grew crops, and made stone tools that have been identified in other settlements two thousand miles away—proof for any in doubt that some system of repeated voyaging existed between these far-flung enclaves. In Tahiti I have heard estimates of up to half a million people, living spread among lush volcanic valleys where the steep topography served as a buffer against intervillage conflict. Between garden plots, coconuts, pigs, and fish, the islands were nearly limitless in their capacity to support life until the catastrophic arrival of Europeans—who, like me, were white men in sailing ships. In truth, nearly every island is a place where people once lived and were doing just fine until the sailors started showing up. We all live in the present we are given, but in my comings and goings from such places, these are the thoughts most difficult to reconcile.

"You are a colonialist," one of my own students will tell me several years later on a stop in the Polynesian Kingdom of Tonga. "You have no business here."

I am just passing through, I think.

"Do you know that Tonga has been self-governing since its beginning, never colonized?" I want to ask them.

Our Tongan friends are waving from the dock, the family of a local scientist who's been an enthusiastic collaborator aboard for the last two weeks. I don't feel particularly colonial, but the questions are more complicated than that. I am not sure what I think beyond my gratitude at having a job that's let me sail to this island and meet the remarkable people who live here. In the end I say nothing, choosing instead the oblique silence that is the time-honored retreat of ship captains. Maybe that makes me part of another problem as well . . .

I recount some of this to my wife in an email.

"I get it," she says. "When we sail from place to place, a challenge is that we are rarely anywhere long enough for our thoughts to resolve—to me they are like bits of laundry hung from the rail. They appear randomly, flap in the wind, never quite dry before the sun sets and it's time to take them back below."

At the grocery store up the road from the gendarmerie are long baguettes for thirty-five cents, a paltry price no doubt guaranteed by some edict of the French socialist utopia. Égalité! As I will soon discover, they are the only cheap thing on the island. It's a crowded little store, and goods spill out onto the concrete porch where dogs sleep like sausages on the steps. Humboldt Current notwithstanding, it has begun to rain buckets, and the townsfolk rattle up in Land Rovers, dash through the torrent, and depart protecting their bread under shirts and umbrellas. There's a guy standing next to me in Quiksilver sandals and a ratty T-shirt. He turns toward me, his face adorned with striking tattoos. He has a ring in his ear, and dark hair pulled back and tied up with a knot of feathers. The rain abates briefly, and he smiles hello, flicks a cigarette aside, and steps out into the street with his bread.

In the afternoon I go walking with Kara and Jeremy, who I find browsing among the mounds of breadfruit and yucca in the waterfront produce market. Married in September, they resemble a movie couple in their matched attractiveness. After several tries we find the service road up to the radio towers above town—a can't-miss scenic vista extolled by someone we meet along the way. From here the view is straight south, at the edge of a serious abyss that drops down through descending terraces of vegetation and ends with water. It's no place for a nap. The *Seamans* is visible far below us in the harbor, drawing gently on her long bight of anchor chain. The island of Ua Pou floats in the distance, an abrupt wedge of green with its top vanished in cloud. One ridge over from town is Colette Bay, where the road goes past the pocket-sized luxury resort with its infinity pool and iced Hinano beer. At the bottom of a mini-valley, someone has a small barn and a patch of bananas. A few horses graze, and otherwise there's nothing. Nothing to tell you that the largest settlement in the archipelago is a mile up the road, or that there's even anyone else on Earth. There's a little gray sand beach where crabs scuttle with claws at port arms. Underwater, stingrays undulate slowly over the smooth bottom, browsing for their supper.

In the twilight calm, canoe crews practice racing strokes in smooth slices across the bay, their boats shiny new renderings of traditional craft—slender main hulls, or *waka*, barely above water and the smaller *ama*, outriggers kept generally to windward for stability. Connecting the two hull structures is a crosspiece called the *aka*, a word that, like the others, I first learned as a teenager from a family friend who built multihulled yachts for ocean racing—hyper-fast competitive machines, their DNA extracted from wood and fiber ancestors that spread a whole society across ten million square miles of ocean. It is Christmas Eve, and in the rapid onset of tropical darkness we walk back past a church, trimmed with extraordinary carved reliefs and bathed in dappled shadow from a surround of trees and garden lighting. This is, I will learn, the Notre Dame Cathedral, built in the 1970s on the site of a much older sanctuary. Christianity is like an introduced plant here, alien but now flourishing inextricably across the landscape. For now all it's possible to notice are the hymns, sung in Marquesan and rolling out from inside the open structure in a polyphonic tide of sound. It's a fifteen-minute walk from here back to the dock, and over that interval nobody says anything, other than to return greetings from families gathered under the low trees along the waterfront.

Cruise ships visit here occasionally, including the *Paul Gauguin*, whose namesake died on nearby Hiva Oa in 1903 after a long period spent making himself unpopular with the locals. In some internet postings from her passengers, I find mention of the cathedral we'd come across in full song. "More of a church than a cathedral," says someone from Las Vegas. "Worth a visit, but not a special trip to the Marquesas."

⌐◦⌐

THERE are amoebas on the radar," says the second mate. This seems worthy of my attention. I step into clogs and start up the ladder to the chartroom, past a line of oilskins swinging gently together on their hooks. We departed Taiohae in the dramatic late-afternoon light of Boxing Day and have been roaring westward ever since, the air growing steadily hotter and damper. Sure enough, on the radar's green screen a faint blurred loop has painted itself into view, a wavy oval ten miles across. Then others, floating like cellular bodies on a microscope slide. Here in the middle of the world's largest ocean there is nothing else to see, other than the great towers of cloud that are suddenly everywhere, dragging dark bands of rain underneath. As the nearest amoeba draws close, there is a slight change to the sea surface nearby, the quavering horizon of wave tops tightening into a dark line distinguishable through binoculars as a single row of palm trees melting in the heat. A cell membrane of earth held tight between sea and sky.

Atolls form as the basaltic cones of ancient volcanoes slump back into the sea and fringing coral reefs race to keep up with their sinking foundations. They are the quintessence of dynamic equilibrium, a standing balance made from moving parts. If the coral is healthy, it may add material fast enough to match pace with the retreating geology, until all that remains to see is a fragmented annulus of breaking waves and *motu*—vegetated sand islets—looped around a central lagoon. At the midpoint of this process the parent mountain may survive as a dramatic pedestal surrounded by its outlying reef, thus opening a window on the relative age of different islands in a chain. Consider Tahiti, which is the youngest of the Society Islands—a mountainous mass thirty miles across and surrounded by only a narrow ribbon of lagoon. To the reef from shore is a swimmable distance, even for tourists. Moving west, the islands get lower and the lagoons larger—Huahine, Raiatea, and then Bora Bora a day's sail away, its alpine spikes of gray stone fringed with jungle and balanced by a dazzling lagoon of almost equal size. Freshwater runoff has cut channels through the reef, building passes through which boats can enter a perfect natural harbor. Farther on, the land grows older still and starts to vanish, leaving only donuts of dry coral with names unfamiliar even to travel readers—Maupiti, Manuae, and Motu One, a fully closed circle whose beaches are left for the terns and coconut crabs.

A mature atoll, mere feet above sea level, is invisible from any real distance—creating for the mariner an opportunity to run aground without ever even seeing land. Their beaches are perched on submerged cliffs—the water a thousand feet deep within yards of shore—meaning that regular soundings might still give no warning of an island's proximity. It is a recipe for shipwreck. Even radar has its limitations. A low beach on a rough night may appear on the screen to be just a dense band of rain, or nothing at all. Seen up close, these attenuated scraps of dry land are barely more than causeways built in mid-ocean—broken paths of scrub and white sand running off to vanishing points in either direction. Sometimes the opposite limb of an island can be spotted in the distance across its lagoon, the water between implausibly blue.

To grasp the life cycle of such landscapes, find a map of the Pacific—preferably a small-scale nautical chart that gives proper shrift to bottom features, what an oceanographer would call *bathymetry*. My first impression of such charts is to think just how little land there is to see at all. Anyone who doubts that our planet is mostly water should view it from this vantage and have their skepticism permanently cured. After all the water, what you may notice next is how all the islands—when they do appear—fall neatly into line, navigating long routes in an arrangement that explains the course of their genesis. The Hawaiian chain begins amid molten pyrotechnics at the eponymous (and geologically brand-new) Big Island and then runs northwest, farther than most people realize—a row of diminishing dots strung nearly to the 180th meridian, halfway to Japan. At Kānemilohaʻi (French Frigate Shoals), five hundred miles from Honolulu, sea turtles loll on a twenty-mile rim of sandspits surrounding a remarkable obelisk of gray stone, the last chance you have to see a real rock until

Tokyo. La Pérouse Pinnacle, this rock is called, named for the French explorer who chanced upon the place one night in 1786—in what must have been a disconcerting call indeed from the lookout.

Warm water and sunlight feed a live geology of organisms, striving toward the surface as the rocky underlayment of earth recedes from underneath. From Kānemilohaʻi it is another eight hundred miles to uninhabited Moku Pāpapa (Kure Atoll), the last dry land in the state of Hawaiʻi, after which all that remains are the Emperor Seamounts—sunken ancestors of islands that were. The movement of oceanic crust over plumes of heat in Earth's mantle is what makes all this happen.

"Think of a thin sheet of rubber," says a geologist I know, "drawn slowly across a candle flame. That's Hawaiʻi."

Carried by plate movement into waters too cold for coral, the fringing reefs eventually lose their capacity to keep up with the sinking mountains, and the atolls are gone. The Darwin Point, this is called.

TODAY's amoebic islands are the Tuamotus, a vast spectral archipelago stretching from near Tahiti all the way east to the Pitcairns—notorious bolt-hole of the *Bounty* mutineers but also the newest rocky extrusions in a continuous land factory. The Pitcairns are themselves future atolls if the requisite epochs go by and balances are maintained. We are at the moment amid the gas giants of this atoll solar system: Rangiroa, largest of the Tuamotus, is 125 miles in circumference, its lagoon spacious enough to hold cruise ships. The tidal currents in Rangiroa's entrance reach 7 knots as its great near-captive sheet of water ebbs and flows, forming standing waves big enough for surfing. "Passage at slack water," say the sailing directions, "is recommended." Rangiroa and its long string of carbonate cousins appear in the earliest accounts of European explorers—starting with Magellan, who in 1521 saw the island of Puka Puka and named it San Pablo, grouping it rather randomly with present-day Flint Island into the Islas Infortunadas. Puka Puka was the first island sighted by the Norwegian anthropologist Thor Heyerdahl (swashbuckling visionary or racist crackpot, you decide . . .) near the end of his long drift across the Pacific in *Kon-Tiki*, a balsa raft he'd built in Peru to prove his dubious theory that the original Polynesians had actually come from South America.

Named the Archipel Dangereux by Louis Antoine de Bougainville after a sail-by in 1768, the Tuamotus were no doubt a terrifying cheval-de-frix to any Western vessel attempting to get past them. At the same time, they were surely a garden path to the islanders, working their way east in small craft more suited to chance encounters with land in mid-ocean. The motu offered a place to pause, replenish, perhaps even stay—though with their thin ground and sparse water supply they were never the equal of the bulkier islands in sustaining settlements. To the east of us is Moruroa, where between 1956 and 1996 the French tested their nuclear weapons—first in midair and then in boreholes drilled into the lagoon floor. In 1979 such a bomb got stuck and detonated halfway down its eight-hundred-meter shaft, leaving the atoll's stone core cracked like a broken tooth. Trapped to leeward, the countries of Oceania watched in dismay, their political leverage for many years insufficient to force change.

When I'm sailing other coasts, the chart edges tend for me to fill with the familiar, even when I'm far from home. San Diego, Miami, western Ireland, and Bermuda, all are places where I can picture scenes ashore and sometimes even read the bus maps in my mind's eye. Here is the opposite: We are surrounded at close hand by amoebas and in the distance are locations even more mysterious. Once-flourishing Mangareva, whose local population dwindled for unknown reasons long ago and later endured decades under the thumb of a mad Jesuit priest named Honoré Laval. Henderson Island, where survivors from the sunken whaleship *Essex* spent just a week scratching for water and falling into holes before most chose to put back to sea in their open boats. Notorious Pitcairn, and farther away still Rapa Nui, where the community set its formidable skills to building the island's signature moai—a thousand unexplained stone statues, some the size of school buses stood on end.

These islands and the seventy or so other Tuamotus host about 15,000 inhabitants now, who in addition to some French speak their own discrete branch of the Polynesian language, Pa'umotu. There are pearl farms, resorts, and local communities sustained by the traditional resources of reef and garden, all pressed between the lagoon's green lens and the open ocean. I have yet to visit very many of these places, but in this aspect they recall to me the words of the author Mark Vanhoenacker—a pilot who writes elegantly of unwalked landscapes sensed instead by overflight. It is a notion he credits to the Alaskans, who may cross broad reaches of their trackless state from above, borne aloft in tiny planes to their own personal corners of the wilderness. I feel this way about the Tuamotus, which for now are like the rings of Saturn passing in the window of my spaceship—unexplored but captivating, if not entirely inviting in their presence.

Clouds boil into the sky around us, atmospheric volcanoes that I suspect are driven by massive evaporation and instability over the warm puddles of lagoon water. Underlit by a fantasy palette of green and blue, they hover above the shallows and then drift offshore, bringing us hourlong lashings of rain, sudden calms followed by lurching gusts that render sailing almost impossible, the timing between torrents too short to set sail productively.

"This is like trying to pitch a tent at the end of an airport runway," I tell the mate.

He regards me as if that particular challenge had never occurred to him. We take in sail and I call David to start the main engine, imagining Bougainville here on a day like this—floating helpless in his wooden ship, hung with creaking spars and frayed hemp canvas. He asked for it, the islanders might say, standing by their canoes hauled safely up on the beach.

HERE in the final third of the voyage, the trainees are for all purposes running the ship. This is the cadet model, the mode of empirical learning that is manifest one way or another in every skill-based training environment. The medical student closes the incision; the apprentice plumber installs a new toilet while the master goes for coffee. A flight instructor directs their pupil to land, keeping one hand discreetly near the control column. Today a young woman named Eliza has charge of the afternoon watch. She appears in the doorway, still wearing her spiked hairdo from the equator crossing two weeks ago. I look up from my copy of Moon's guide to French Polynesia. Our conversation begins, at the pace of lunch being ordered in a strange country.

"Hi, Elliot."

"Hi, Lizzy. What can I do for you?"

"Well, there is a boat nearby, about two miles away."

"OK . . ."

"And, um, they are waving their arms at us and shooting off flares."

Well, then. This is a nonroutine event. We are leaving the glide path. Jeremy, the flight instructor, appears behind Eliza, and with a smile and muted hand gesture suggests she turn sideways—as though opening a door that I hurry through on my way to more information.

The motor vessel *Tehinanaki III*, of Tahiti, is rolling deeply in the swell as we come abeam. They have put away their flares at this point, but I notice the strobe of their rescue beacon flashing briskly from the cabin trunk. She is a small sportfisherman, about thirty-five feet long, of a type that would not look out of place in Miami or Montauk. The men in her cockpit lean on fuel drums and watch us calmly. Nobody answers the radio. Perhaps it is dead. They are, we will soon learn, on their way to Nuku Hiva for a family visit. Their fuel stores, normally designed for a day of inshore fishing, are buttressed by a deck cargo of diesel, but through some oversight they have let their main tank run dry. Restarting a diesel engine in such circumstances requires first bleeding out the air it has inhaled—a job involving tools and technical acumen that is constrained to a limited number of attempts before the starting battery is exhausted. It is at this extremis that the holiday voyagers have arrived—their fuel tanks refilled but engines still dead and battery drained. Somehow their water tanks have been emptied also, a development that is never fully explained.

Compelled by goodwill and the seafarer's ancient obligation to render assistance, we set to finding a solution. David and an ad hoc translator, found at random from among the students, are bundled into our Zodiac inflatable after a brief shouted conversation. They heave alongside us to get ready, David adjusting his stance like a surfer to handle gear as it is loaded. Toolbox. Water jug. A spare battery, filled with acid and brutally heavy, perhaps first among things you'd rather not carry with you in a small rubber boat. With Jeremy at the tiller they speed away, vanishing at intervals in the long relief of waves. I am watching from the rail, my hand raised against the low-angled late-afternoon sun.

Kara appears, visibly disconcerted. Her new husband now a castaway.

"Can't we just leave David behind with them?" she asks.

Three boats roll near one another in the swell, daylight waning. In the great clouds sailing away from the lagoons a billowing texture becomes visible, white monochromes replaced by the creamy palette of twilight. None of the squalls threaten us until almost dark, just when from a sudden puff of smoke and roar of noise I realize that David has succeeded in his mission to get the Tahitians going again. His voice soars over the radio to confirm this in an exultant shouted exclamation, its precise wording unclear. Jeremy runs the boat back to pick up our team just as the world disappears in a dark blast of rain and the concurrent end of daylight.

We are drifting without sail, but the bystanders and loose equipment that have accumulated around our rescue effort disperse like a crowd hit with a fire hose. A quiet afternoon is randomized into chaos, our boat floating out there somewhere in the maelstrom with our chief mate, chief engineer, and a student—who, thanks to minoring in French, is now having the adventure of his life. Behind me the steward's assistant walks past, ringing the bell for dinner. The routines of a ship are indelible, I am reminded, moving parts persistent in their channels unless directly disrupted. It is the strength and the vulnerability of such systems, a movable key to safely addressing the unexpected. There is no "rescue the Tahitians" column in the station bill, no cell printed in the matrix to advise the galley not to ring for dinner in the middle of a squall. The band plays on.

Water is running down in a dribbling stream from a badly closed hatch in the chartroom. The radar is overrun by a green blob of precipitation that inches past as I watch, fiddling with knobs to see if I can find either of the two nearby boats on the screen. There is nothing, and then a sudden call from the deck to tell me that the rain has stopped. Sure enough, it has, and I can see the luminous bow wave of our dinghy making its way back to us, the beam of a flashlight tipping up and down. The Tahitians are just beyond, their own lights now illuminated and a steady rumble from their engines, a sudden burst of music from their deck speakers. They have wasted no time in returning to full operating status. Quickly they also are alongside, four large men in a small boat, lurching on the swell. Two in deck chairs, a third balanced casually near the rail to hand back our forgotten water jug. The captain, calm as a bus driver, feathering his throttles to keep close without touching as their little craft lunges next to us.

There is a handoff and a shouted appreciation:

"Merci, amis! Bonne chance!"

And then they are gone, lights and music vanished in the dark. They have about six hundred miles left to go, a distance that I suspect gives them no pause. They are not the first Tahitians to set out as they have in a tiny boat. I think of Greenland, and how during my trips there I grew used to hearing faint engine sounds in the quiet vastness of an Arctic bay—an echo resolving gradually into a small skiff whose occupants, an Inuit family or a single hunter in a plaid shirt, would wave a casual greeting, passing in a Doppler burst of noise and vanishing back into the landscape.

Our boat team stows their gear and sits down with plates of leftover dinner. "A rescue at sea!" says someone. Jeremy nods. His time in the Coast Guard included scarier situations, their stakes higher and solutions more tenuous. On tour as a new lieutenant he led a team aboard a freighter suspected of carrying drugs, standing guard with a sidearm on the bridge as his search party made their way through the ship. The terrified helmsman, perhaps hoping for leniency, blurted out a secret.

"The cocaine," he said, "is hidden in the mast."

This turned out to be true, a source of brief triumph to Jeremy's team until they realized slowly that the ship—scuttled by its crew upon the discovery of their contraband—was in the process of sinking.

In addition to its law enforcement mission, the US Coast Guard can do things like fly through hurricanes and hoist victims to safety in unfathomable conditions. This said, the majority of rescues at sea are effected by regular civilian vessels, passing by chance or diverted from their routes by rescue agencies unable to send assets to every potential casualty. When the sailing school ship *Concordia* sank off South America in 2010, her crew were rescued by the Japanese wood-chip carriers *Hokuetsu Delight* and *Crystal Pioneer*, at the behest of Brazilian authorities. The masters of these vessels brought their ships directly alongside the life rafts to perform the recovery, maneuvering nine-hundred-foot cliffs of steel as though collecting a water skier.

Nine years after the *Concordia* rescue, the *Crystal Pioneer* passes us by chance one day north of Auckland as we are performing an abandon-ship drill.

"Calling *Robert C. Seamans*, this is *Crystal Pioneer*, ZCYA."

"Yes, *Crystal Pioneer*, this is the *Robert C. Seamans*, WDA4486. Go ahead?"

"Yes, good afternoon, *Robert C. Seamans*. You are every-thing OK?"

The sea limitless, yet at the same time oddly small.

A T the entrance of Tahiti's main harbor are the first naviga-tional markers we've seen since Mexico, a neatly painted pair of buoys in the roiled waters of the pass, where a steady outflow is forced by the return from waves breaking over the reef. The current can surprise you, I was warned, especially if there's a big sea running and the lagoon is high as a consequence. The reef is several hundred yards offshore, a perfect natural breakwater but pancake flat, empty of cues for guidance. From afar the port itself is invisible, flattened against the green bulk of land until at a distance of about a mile things begin to resolve. Previous to this there is the feel of rushing in at a blank wall of mountains, with signs of development along the shore but little in evidence of a harbor. Two small structures—what mariners call a *navigational range*—are positioned on the hillside to help keep your faith. If the low tower in the foreground appears lined up with the taller one behind, you are yourself in the middle of the channel. Our ship has not been near a dock in six weeks. At times like this I wish pointlessly for some sort of buffering period, a gentler introduction to hard surfaces after so long in the open.

A string of ferries is on its way out the channel, and the harbor controller is calling us, asking to know our intentions.

"Good morning, *Robert C. Seamans!* You are at what time expecting to enter?"

Tahiti has built an international airport on top of its fringing coral reef, likely an option that many architects would choose if they were able. Imagine billions of tons of natural concrete, delivered for free and poured perfectly level in a long straight line just parallel to shore. The only drawback here is the ship channel, crossing just east of the runway, where a poorly timed arrival could trigger the worst of intermodal disasters. This is the key to our permission, and with the morning flight just departed we are authorized to proceed. The radio operator is positively cordial.

"*Robert C. Seamans*, you are cleared to enter. Bienvenue à Tahiti!" he says, a maître d' welcoming us with all the contentment of a civil servant in paradise. Water pours out of the reef cut, and there is someone surfing the break east of the channel as we go by. I am watching the range markers and giving instructions to the helmsperson, our world after all this time suddenly smaller and much faster in passing. A ship is turning around in front of us, preparing to leave. We run up the channel, nearly to a beach where long rows of stacked canoes are clearly visible, color by color like crayons in a box.

"Why did we leave that green buoy on our right?" someone asks.

"Let me get back to you about that," I offer.

Next comes a turn to our left, nearly 90 degrees, and toward the berth we've been assigned, which is at the top of a long basin past a huge yacht, its paint as deeply lustrous as a swimming pool. David is at the engine controls.

"Slow ahead," I tell him.

We make our way toward the dock, the large yacht now drawing up alongside us, our image reflected back from her topsides in rippling flickers. There is someone on the foredeck polishing railings with a towel, and a young woman in very white shorts carrying a tray down a ladder who stops to watch us go by. On the wharf ahead of us is a bulky Tahitian in a hard hat and Day-Glo vest, standing in the universally indolent way of dockworkers worldwide. Next to him is our agent, equally unmistakable in his own standard kit—polo shirt, sunglasses, a folio of important papers under one arm.

We are nearing the top of a blind alley, the concrete wharf blocking our path at an apparent right angle—actually canted ever so slightly away from us, according to its rendering on the harbor plan. There is a tricky surge here, I've been told, a sort of sloshing effect in the lagoon driven by the constant overtopping and recession of ocean waves. I now can see evidence of it along the wharf bulkhead, a slow undulation of the calm surface, the slight lifting of a debris slick alongside us. I smell the earthy aroma of tropical vegetation, mixed with French fries and car exhaust from the boulevard.

"Stop engines."

We are still moving ahead, just barely fast enough to steer, the flawless paintwork of the yacht at seemingly touchable distance. I look aloft at our yards—long horizontal spars, their tips studded with hardware—braced sharply around at an angle to make us as slender as possible. The man with the towel has stopped polishing to lean on his rail and watch. We need to stop soon—actually, to stop while turning, controlling our momentum long enough to end parallel to the concrete wall now dead ahead. I think of yoga.

"Slow astern," I offer. "Fifteen degrees right rudder."

A single-screw powerboat put into reverse takes considerable time to start going backwards, though it will almost immediately yaw to the right as the rotating propeller wash interacts with the hull. I feel the propeller bite. For an instant the yacht and its million-dollar paint job are breathtakingly near, then tailing away as our bow swings and progress carries us away at an angle, now parallel to the wharf face.

"Stop engines."

We have floated to a halt, nearly in Tahiti, gazing across four feet of murky water at our line handler. A surge fills the basin; we rise and settle and, as though planned, slip neatly sideways across the remaining distance, our rubber fenders absorbing the not-inconsequential impact. We are there.

"Tie up all lines," I say into the melee of bulky cordage that, up until an hour ago, had been sleeping peacefully in a locker all the way from Puerto Vallarta.

Last in the order of steps is to confirm with David that we're done with the engine, and to let the helmsperson know we are finished also, that they may secure the wheel to midships. For difficult dockings like this one I typically choose an experienced member of the crew to steer, but in this instant I realize I have forgotten. At the helm is Eliza, who throughout the high-stakes game has not missed a beat and now looks back, eyes like dinner plates, when I say her name.

6

THE MOTHER SHIP

The Center for Weather and Climate Prediction sits like a grounded starship on a backlot at the University of Maryland campus in College Park. It is a striking product of the post-rectangular school of public architecture, its entrance well-concealed from the inexperienced pedestrian. When I do get there after several missed approaches, some guards at a lobby desk wave me in, their gestures barely visible through mirrored glass. The doors are twice my height. One of them buzzes at me and I fumble with the handle, searching my pockets for the ID I will need when I get inside.

My host is waiting for me at the metal detector. Joe Sienkiewicz left his job as a tugboat captain to become a meteorologist in the mid-1980s, right around the time my own career as a sailor was beginning. I first saw his name in 1990, when amid a fraught breakup with my college sweetheart I joined a schooner bound across the Atlantic for Portugal. It was late October, the heart of gale season, and as I suffered with my shipmates through an unrelieved month of storms, our punishment was foretold each day on blurred NOAA weather maps, curling slowly out of a fax machine. Every such map is signed by its author—a convention that gives unique accountability to these composites of raw data and the forecaster's art—and there among other names now etched in my memory was Joe's. From my remote vantage the forecasters took on a mystic quality, not unlike the masterful artists who might ink their names into the final panels of a graphic novel.

Today Joe is a branch chief at the Ocean Prediction Center, one of eight units based aboard the meteorological leviathan that I have just boarded. His office deals with ocean weather on the American side of the world, in tandem with two similar centers in Honolulu and Miami. A quarter planet's worth of water. Their collective bailiwick is deep-sea meteorology, the stuff of ships on offshore voyages. Forecasts for coastal waters—what you might consult before an afternoon of bass fishing—are handled by a cloud of smaller offices spread around the continent. There is one such outfit not far from where I live in Maine. When I contacted them looking for a tour, they hung up the figurative phone. Sorry, they said, too busy. Fair enough. To my great good fortune, Joe was more than happy to oblige, blessed perhaps with a more forgiving schedule and what he describes as a devout interest in the welfare of sailing ships.

"They are slow moving and weather sensitive," he is fond of saying, "and carry precious cargo."

The building's core is a soaring atrium, overlooked by balconies and lit through louvered glass exterior walls. I look around carefully for anyone who might be carrying a light saber. As we climb the stairs between each floor, placards mark the doorways of other departments: The Climate Prediction Center. The Environmental Modeling Center. The Center for Satellite Applications and Research. The OPC rooms are on the upper floor, and when we arrive, it's just after eight a.m., Eastern Daylight Time.

That's noon in Coordinated Universal Time (UTC), one of four main benchmarks in the daily forecast cycle. The center has about twenty meteorologists, working five at a time in ten-hour shifts. They're all up at one end of the main operations area, a long airy space dotted with workstations. People tap away in deep focus near screens splashed with exotic displays. It's surprisingly quiet. We pass groups from other divisions who all pause in their work to nod hello. There's a greeting in Spanish from a section of South American scientists, all of whom are dressed several points up the fashion scale from their nearby counterparts.

Frank Musonda is camped behind five wide screens in the OPC cluster, just getting started on the Pacific surface analysis—a snapshot of the atmosphere's very recent past that is rendered four times each day. He invites me to watch. Frank starts on the noon map at 1230 UTC, enough time to ensure that data have arrived from all of the observing ships spread across his jurisdiction. He begins his composition with a basic overlay of pressure, temperature, and wind—digital graphics generated by the various weather models running in a constant hum on supercomputers around the world. Forecasters can superimpose data from different models and decide which fits the best when measured against actual observations. The American model, called GFS, and the ECMWF, from Europe, are often close but rarely in full agreement. Products from other agencies may give better local resolution in certain cases.

Today is an average spring day in the North Pacific, with the trade winds well-developed around Hawai'i and several low-pressure systems working their way from west to east above the 30th parallel. My friends in Alaska are being sadly pounded with wind and rain. Frank lets his software draw the isobars—lines of equal pressure—and makes adjustments by hand where they are needed. Then he paints in frontal boundaries and surface features, their borders based on his own reckonings and a stream of analytic data that helps him identify key breakpoints in temperature and moisture. Here the qualitative weight of a forecaster's experience is added to the swarm of ones and zeros, a live estimate of what is actually happening down there on the water. To help with this there are satellite images, along with the critical recordings from ships at sea, each discernable by its alphanumeric radio call sign on the image. Sure enough, there is the *Robert C. Seamans*—WDA4486—not far south of Oahu, its winds shown by an arrow pointing resolutely from the east. If I think carefully, I can identify the captain who is aboard in my stead, standing somewhere with a coffee cup or gazing idly into the radar. On reflection I realize it is not long past midnight in Hawai'i, meaning that whoever my counterpart is, they are likely asleep.

Frank's mandate ends at the continental margins, and in finishing his first draft he scans maps from other offices to ensure that what he has drawn will mesh easily at the boundaries. He likes working the Pacific because the weather there is what he calls "smooth"—far from land, it has found its unchecked pace, like a horse free from the gate and well down the racetrack. Frank's first deadline is at sixty minutes past the noon hour, when he must send his product to the partner offices responsible for neighboring maps. The shared images are passed back and forth to ensure that there are no discrepancies in what's shown. The different forecasters speak to one another in shorthand text that pops up in a dialogue box on-screen:

TAFB [Miami] is waiting
OPC is ready
TAFB has received
OK
HFO [Honolulu] is good

The map takes shape, human hands creating in real time a thing that I have known for most of my career as a sacred object. It is like watching money being printed. NOAA has its own data servers, Frank tells me, forming what is effectively a private internet. This is important because it is the last place you want hackers messing around. He is originally from Zambia and speaks with a diction suitable for lecturing at Oxford.

"You need to think of this as writing a history book," he says. "Once you send it out, it will be read by thousands of people you don't know. And *your* name is on the map!"

The forecasters work quickly, aiming to stay ahead of their deadlines as a courtesy to one another. Frank must have his map finalized in time to send out before 1519 UTC, the slot for its scheduled broadcast to ships. If that doesn't happen, the map will be "junk"—available over the internet but disqualified from the stream of weather that is transmitted by Coast Guard radio stations. His final task is to draft a text forecast, effectively a written description of everything he has just drawn. This is a slow and artful chore—just a half step ahead of what people like Joe learned years ago as junior forecasters, drawing with grease pencils onto sheets of acetate.

The science of forecasting now walks a wavy line between human judgment and an unfiltered acceptance of what our instruments and modeling programs are telling us. In the app-driven age of information, rapid changes are at play. The US National Weather Service employs several thousand forecasters. The Weather Channel employs just under two hundred. Some forecast applications use none at all, relying solely on the gridded stores of raw model output to create their products. Click your screen, get the weather anywhere in the world without pesky human interference.

Currently the OPC does not disseminate most of its information this way, though the future is leaning away from handwritten messages. Joe's office is working toward an expanded suite of digital products—including maps that can be mined for point-specific information and forecasts that use graphics rather than text to spell out the probability of certain occurrences over time. One challenge to applying these improvements in the maritime world is the need at each step for a painstaking revision of international standards—which at present still require forecasts that can be read by a human voice over the radio. Another constraint is bandwidth. High-speed internet is on its way to sea but hasn't yet fully arrived. Large ships now have the capacity to access data-rich products via satellite from the web, but many smaller operators still do not, particularly in offshore waters.

Many commercial ships employ weather routing services, private forecasting firms that advise them on where to go. Even now, much of that advice is still based on climatology—that is to say, historical rules of thumb. Container ships may elect to cross the North Pacific along the 30th parallel of latitude because by the odds that's the safest place for them to go—north of the tropical cyclone tracks, and mostly south of the subarctic storms that can break boxes free of their lashings and toss them overboard like unsecured deck chairs. For some vessels this means going hundreds of miles out of the way, with millions burned annually in excess fuel. Sometimes even this precaution is not enough to avoid catastrophic losses of cargo.

Think, says Joe, of how much better they could do with a complete stream of dynamically derived data, available in real time. I do think about this. One can marvel at how much more we know about weather now than we did in 1900—when a poorly forecast hurricane came ashore in Texas and killed ten thousand people— or might look to February of 2016, when the cruise ship *Anthem of the Seas* steamed willfully into what turned out to be a horrendous winter storm off Cape Hatteras. There were no deaths or serious injuries suffered in the beating that followed, but this without doubt was an example of how much there still is to learn. Joe pulls up slides from a presentation he gave on the *Anthem* incident. The brand-new ship sailed south from New York just as a storm was forming north of the Bahamas, a classic "bomb cyclone" that deepened explosively as it crossed into the warm Gulf Stream. Off the Carolina coast, the *Anthem* experienced 145-knot winds from the northwest, just aft of her starboard beam. Unable to hold course or turn into the weather, she blew sideways at 10 knots, rolling like a dinghy while furniture and potted palms swept across her interior in mini-avalanches. Four thousand passengers were confined to their cabins for safety, shaken like airport baggage. Those who had windows took pictures of waves higher than the main deck.

Forecasters still want for sufficient tools to predict such explosions consistently. Joe and I look at animated satellite images of the *Anthem* storm, which was expected to produce 65-knot winds but delivered more than double that in reality. The clouds are a vortex of meringue, spinning at what seems an implausible speed for mere weather. They are in the midst of an intensification that even Joe cannot fully qualify. From far above, the system has the look of a twisted hammerhead, a great trailing hook of clouds circling an open area behind the center. Underneath the clear spots it's possible to see the Gulf Stream itself, shadowy gray in the thermal imaging.

Joe stops the video and shakes his head.

"There is so much going on out there," he says, "that we still don't know about."

The crew of the *Anthem*, rich in experience and surrounded by information, were blindsided. Through gleaming bridge windows, their propellers overmatched, they watched the arrival of the unimaginable. On innumerable days in the future, from my own exposed deck with rain running down my glasses, I will share their dismay. One can study these things indefinitely, but the surprises never really end.

~~~

THE first marine weather maps were drawn more than 150 years ago at the new Royal Meteorological Service by then-Admiral Robert FitzRoy and a team of colleagues. Building on the work that William Redfield had done with hurricanes in America, FitzRoy produced the first full view of an extratropical cyclone in a rendering of the catastrophic Royal Charter Storm of October 1859. The Meteorological Office had been working on its new so-called *synoptic* maps for several years, but the Royal Charter effort was their inaugural tour de force. Through a web of collected data points they drew out the story of this devastating late-summer storm that had closed its jaws like a pit bull on the British Isles. Reports from ships and far-flung coast stations allowed FitzRoy to paint his seminal picture of two great air masses caught in shear: a run of fine days ending in a murky pall of fog, and then in the dark of night a sudden brutal wind from the north—gusting over 100 knots, destroying barns, roads, and entire fishing fleets with awful equanimity. At the center of it all was the steam clipper *Royal Charter*, attempting to complete the final hours of a sixty-day passage from Melbourne to Liverpool. Having declined the option of a last-minute stop at the Welsh port of Holyhead, she was driven ashore and wrecked on the Isle of Anglesey with the loss of some 450 lives—more than 90 percent of her complement.

This tragedy propelled a campaign to build a network of signal stations for storm warnings along the British coast, each charged with hoisting certain markers when reports indicated an approaching storm. Some station-keeper in the west of Ireland would observe a telling wind shift or change in the barometer, a horse-man would gallop to the nearest telegraph, and then FitzRoy and his team would set to work and decide if warnings were warranted. It was all very in-the-moment, constrained by the range of human vision in the mid-nineteenth century—though similar tools remained in use eighty-five years later, when weather reports radioed from western Ireland were a key final piece in the decision to proceed with Allied landings on D-Day.

⌒

I N the William Gibson novel *All Tomorrow's Parties*, a character bemoans the ad hoc evolution of the microprocessor, noting how different the computer would look if it could only be designed from scratch. The National Weather Service today has a similar feel, a machine knit together from small parts that's now tasked with a much larger job. Weather, like politics, is at once local and all-encompassing. Seen on paper, the service is an awesome bureaucracy to behold, though it is still just one branch of NOAA, the National Oceanographic and Atmospheric Administration. NOAA engulfs five other operational offices, including the National Marine Fisheries Service and the Office of Marine and Aviation Operations—a miniature data-gathering navy with white ships, blue airplanes, and no guns. Few people know that it's actually possible to serve as a commissioned officer in the NOAA Corps, just as one might in the Marines. The Weather Service alone has a dozen discrete subdivisions, all overlapping to varying degrees of seamlessness. With the Ocean Prediction Center, there are the National Hurricane Center, the Storm Prediction Center, and the Tsunami Warning Center. There is an Aviation Weather Center and a Spaceflight Meteorology Group, dormant since the last space shuttle flight and not to be confused with the Space Weather Prediction Center. Thirteen River Forecast Centers monitor the state of inland waterways. One hundred twenty-two regional forecast offices—crown jewels of their respective congressional districts—take their own observations, integrate them with global data, and reduce them to a grain size that's useful on a local basis: planning a camping weekend, say, or setting out to haul lobster traps.

The actual forecasts flow from a list of outlets that grows by the day—print, radio, television, and the bright screens of uncounted personal devices. Collectively the NOAA sites are a boggling mobius of information, where even the experienced may fail to spot key nuggets as they roll past in bulleted lists. In the aftermath of Hurricane Sandy's brutal strike on metropolitan New York in 2012, a long series of hearings took place to establish which parts of the warning system had worked and which had not. By all measures, Sandy was not a simple meteorological event. It began as a hurricane and hybridized into a complex extratropical system somewhere off North Carolina, creating immediate territorial challenges within the forecast and warning network. Local responders found the stream of information to be dense but not always contiguous. At a post-event forum one such stakeholder recorded his thoughts on the functionality of NOAA's website for the record.

"Too many clicks," he said.

National Weather Service data are collected with tax dollars and considered to be in the public domain. Most private weather outfits are thus informed by NOAA, even if they employ their own meteorologists. In the strange times of Trump's America, Barry Lee Myers, the CEO of a large for-profit forecasting corporation called AccuWeather, was nominated to serve as the new head of NOAA. He is a lawyer by trade, not a meteorologist. Bizarre as this seems, it's not the first time the agency has been a focus of the tension between public service and private enterprise.

In 1983 President Ronald Reagan's NOAA administrator suggested privatizing the agency's satellites and cutting its workforce by a third. Anyone who wanted the data could buy it back from the firms who had purchased the satellites. None of this ultimately transpired, but the idea remained alive enough to surface two decades later, this time in a failed Senate bill called the National Weather Service Duties Act of 2005. This bill proposed to limit NOAA's license to disseminate free weather data, with the stated aim of promoting growth in the private forecasting sector.

The Weather Service began in 1870 as the Weather Bureau, a somewhat troubled adjunct to the US Army Signal Corps. Forecasting relies more than anything else on the ability to share data from multiple places in some synchronized way—that is, in order to understand what's coming to your own town, you need to know what's happening in the others spread across the map around you. This first became possible with the growth of telegraph around 1850 and was given a second serious boost by the arrival of radio near the turn of the twentieth century. In 1912 the capacity for ships to report weather from sea by wireless was introduced, a linchpin of data collection that remains in place today.

The construction of an effective processing network was also critical, given the immense hunger of this new field for information. Computer models do much of this now, but to start with it was just people—all recording data, sharing it along, and relying on one another to interpret it correctly. The web of early forecasting was not unlike a sprawling hospital with thousands of doctors working furiously in different wings to diagnose one very large patient. Anyone who wants a sense of the service's early travails should read a book called *Isaac's Storm*, by Erik Larson. This is the story of the hurricane that came ashore on September 8, 1900, and reduced the boomtown of Galveston, Texas, to driftwood in an afternoon. Ten thousand people died, mostly drowned in their houses or bludgeoned to death by flying debris. Despite later claims to the contrary, no proper hurricane warnings were issued by the Weather Bureau that day.

This was the Gilded Age, another high-water mark in American hubris when we thought we had things figured out but hadn't really. In their limited understanding of tropical cyclones, Yankee scientists in 1900 were convinced that hurricanes held to a narrow track between the West Indies and the North Atlantic and thus didn't enter the Gulf of Mexico. They also had no real knowledge of the *loop current,* an oxbow of warm water flowing north from the Yucatán Channel, whose heat can ignite small disturbances into conflagrations when they reach the Gulf—much like an oily rag thrown onto the embers of a campfire. The Cubans had other ideas. As natives of the tropics, they possessed a more nuanced picture of hurricanes, but their warnings were ignored. In fact, on the fateful morning that meteorologists in Havana noted the passage of a cyclone and concluded that it was headed for Texas, they were barred by American authorities from sending a forecast to the mainland over the local telegraphy equipment—which in the aftermath of the Spanish-American War was under US control. The Isaac of Larson's title is Dr. Isaac Cline, who at the time was chief of Galveston's Weather Bureau office. In Chapter One we meet him driving his sulky along the beach, watching with a deep sense of foreboding as a long swell begins to break on the shore.

World War I and the boom years afterwards showed how critical weather would be to the enterprises of a new century: artillery, aviation, naval warfare, automobile and steamship travel. By 1919 the bureau was collecting data from the atmosphere with balloons, and providing specialized forecast products for aircraft, highway travel, and shipping. A marine division, the ancestor to Joe and Frank's office, was established in 1920. True to its agrarian roots, the bureau also retained its cranberry warning service, potato frost bulletin, and—as a favor to the honey industry—a special notice to beekeepers.

The next watershed came around 1925, as the department sought to assimilate the insights of the new Norwegian School of Meteorology into its forecasting methods. The Norwegians, led by physicist Vilhelm Bjerknes (father of Jacob), correctly posited that weather was governed by the hydrodynamic interaction of *air masses*—discrete blocks of atmosphere that largely held on to their physical characteristics of temperature and humidity, interacting through features at their boundaries, or *fronts*. A new school of dynamic measurement fought to displace a guild of intuition. Many of the moving parts were still waiting to be worked out, but all of this meant a fundamental change from the descriptive to the analytical in how forecasts were created.

It was during this time that Carl-Gustaf Rossby emigrated from Sweden to serve as a research associate at the bureau. A rising giant who would become the star architect of meteorology in America, the charismatic Rossby lasted long enough on the job to get noticed but was cashiered in 1927 for giving unauthorized guidance to Charles Lindbergh for one of his long-distance flight attempts. Rossby soon found more appropriate quarters in academia, where with a team of compatriots he led the evolution of meteorology into a quantitative science. Much of the support for this process was military, driven by the rapid growth of aviation and fueled by Rossby's insistence that modern forecasters needed a broad and adaptable skill set—a "portable, placeless knowledge"—derived from first principles, not local savvy built by long observation.

World War II was the tipping point in Rossby's campaign for a scientifically based forecasting system, as the operation of modern battle systems—particularly aircraft—relied inextricably on a quantitative means of predicting the weather. The training programs Rossby and his colleagues built during the war provided the critical mass of meteorologists necessary to carry the science forward into the second half of the twentieth century. There were also a lot of good used airplanes around to repurpose for science after the fighting ended, and the pressing agenda of the new Cold War—which relied heavily on aeronautics to maintain its gruesome balance of power.

The world of weather science would spend the next half century evolving into the network we are familiar with today, in an evolution hitched directly to the exploding human capacity to gather and share information. In his excellent 2019 book, *The Weather Machine*, Andrew Blum reminds us that the globalization of weather science marks one of the few unqualified diplomatic successes of the postwar era. Cold War aside, the daunting scope of the atmosphere—coupled with a general recognition that what goes around comes around—has led to six decades of international collaboration at the World Meteorological Organization, a division of the UN. Today weather data are shared among nations, more or less freely.

⌒

NEAR the station where Frank is working, a broad bank of screens flickers with satellite imagery, beamed down from an awesome assemblage of wizardry flying in silent orbit far overhead. A technique called *scatterometry* uses radar to measure wind from space by sensing the texture of the sea surface. The data are compiled into beautiful color maps, like Impressionist paintings of the marine wind field. FitzRoy would weep. This science was pioneered by US agencies, but the workhorse instruments now fly mostly aboard spacecraft from Europe—exotic birds in polar orbits that let them view the whole planet, orange-peel style, in fourteen daily swaths as Earth rotates. NOAA's own Geostationary

Operational Environmental Satellites (GOES) do just the opposite, looking down continuously on a single sector of Earth's surface from an altitude of 26,000 miles—a point where their orbital motion is matched exactly by Earth's rotation. Cameras aboard the GOES units take images of cloud formations and can track the movement of airborne water vapor traveling in atmospheric rivers to deliver rain at the surface. Tropical forecasters rely on GOES data to assess hurricanes, using a method called the Dvorak technique to compare temperatures between the warm eye of a storm and the colder surrounding cloud tops. There's a new set of instruments to record lightning, their imagery shown flashing up on monitors near where Frank is working—the energy of storms sent into space and back again, to this place of carpeted floors and shirtsleeves where the weather feels like it's everywhere and yet somehow quite far away.

It is hard to overstate how profoundly pictures taken from space have advanced our understanding of weather—starting with the ability to simply see the shape and size of each feature in real time. When my father was a teenager, the imaging of storms had more than a little in common with the old fable involving a group of blind men and an elephant—you grab the trunk, I'll pull the tail, and we'll talk about the animal we have hold of. This said, the surface of Earth covers an area of two hundred million square miles, and NOAA—when last I looked—operates a mere sixteen satellites. Both are numbers to remind us of how loose the mesh remains in our web of seeing. Just ask the people aboard *Anthem of the Seas*.

⁓

VILHELM Bjerknes was credited in 1913 with being the first scientist to reduce atmospheric dynamics into a set of workable equations, but in that era of math by hand, the thought of actually applying them defied reason. In 1920 a British physicist named Lewis Fry Richardson proposed the fantastical solution of filling a football stadium with thousands of human computers—64,000, to be exact—all working equations simultaneously and communicating with a conductor via hand and flashlight signals. This is in effect what a modern computer does, albeit with silicon chips instead of scriveners. My own sense of the raw energy involved in the running of weather models crystallized during a tour at New Zealand's National Institute of Water and Atmospheric Research several years ago. Their supercomputers are housed in a building the size of a three-bay garage, just south of downtown Wellington along the shore of Lyall Bay. Our guide gathered us in the parking area, from where you could see penguins diving and surfacing along the harbor edge. It was a warm summer day, and—rare for Wellington—nearly windless.

"We'll talk outside," he said. "In the building it's too noisy, ye can't hear y'self think."

"Noisy?" I asked. I tended to think of computers as silent.

True, I was told, computers can be quiet, but what they do make is heat, enough to require substantial cooling for machines this powerful to run without melting. Hence the argument for placing an asset like this at the edge of a harbor, even if it is in the heart of an earthquake zone. The building's isoelastic foundation will theoretically keep it from sliding into the sea, and in the meantime there's an ample supply of cold seawater for the heat exchangers. We went inside. Our host was right: The pumps were loud as hell. Underneath a long room full of what looked like big black refrigerators was a low labyrinth of howling motors and tangled pipe. I'm a stranger to supercomputing facilities, but as a veteran of ship engine rooms, I felt right at home there. The black refrigerators above us—actually called *racks*—were impressive mostly for the power and expense they so obviously entailed. The technician opened a door, inside of which were row on row of individual computing nodes. They looked like oversized printer-ink cartridges.

"See that, mate?" he said. "One of those costs a hundred thousand dollars."

Cost and noise aside, the other remarkable part of our visit was its congeniality, as our mixed gaggle of sailors and scientists was led back out into the brightness of day, toward the café for pies and lemonade. This is one of two supercomputers in New Zealand, and by far the largest—a keystone of the national data infrastructure. A sandy-haired bloke with sandals and pushed-back sunglasses held a door open for stragglers and smiled in welcome. At home, I thought, we'd still be having our retinas scanned while guards hovered nearby in their flak jackets.

This was in 2017. My next lesson in the pace of modern computing came two years later, when I revisited the NIWA website for more background on the machines we'd been to see. They were gone, replaced already with two even bigger and faster computers named Maui and Mahuika. In photographs they appear nearly identical, though the cabinets, once a faceless black, are now adorned with splashy graphics of indigenous artwork. The miracle machines that we had visited, names unknown, were cooling their circuits on a used computer lot somewhere or perhaps being recycled into toasters.

A MATHEMATICAL model is really nothing more than a group of equations strung together to simulate a process. You could design a model to predict your monthly spending on dog food by figuring the cost per cup and multiplying by the number of cups your dog eats at each meal. You might be doing something like this already, since such simple models inform the lives of even the most math-averse among us. The arithmetic would get trickier if food prices fluctuated, your dog ate less on hot days, or a friend fed him extra breakfast on random weekends. This is why models built to predict complex systems—like weather or the stock market—get complicated so quickly. It's also why we've had to wait until the age of supercomputing for them to grow into their own. NOAA's forecast baseline is commonly drawn from something called the Global Forecast System (GFS) model, which—as the name suggests—is a global simulation that takes a ridiculous amount of computing power to run. You could perhaps start the math with a pencil, but by the time your grandchildren finished the work, whatever weather was coming would be long over.

In truth, your grandchildren wouldn't even get close to finishing. As I was recently reminded, the GFS model requires about ninety minutes to run on its new supercomputer—a machine capable of 8.4 *petaflops*, or 8.4 quadrillion operations per second. For scale, simply counting out loud to a quadrillion would take you about twenty-five million years. In my research voyage to verify these outlandish claims, I ran across yet another interesting factoid, which is that the human brain remains by far the most efficient computer on the planet. There is no contest. Our gray matter is capable of an equivalent of 100 petaflops, all on the energy from a donut or a bowl of oatmeal. Maybe we could indeed do the math by hand, if only our pencils were faster.

My friend Greg is an oceanographer who trained at MIT and now teaches at the University of Maine. Our time together is spent mostly on kayaking, but on one of our recent river trips I pushed him toward shop talk.

"Tell me how atmospheric models work," I asked him.

Like any professional scientist, Greg offered back a fully formed explanation, suitable for publication—this despite the fact that he was wearing a drysuit, digging hopefully into our lunch bag for a stray Clif Bar.

The basic idea is that models cut up some part of the environment into small boxes in 3-D, said Greg. Each box could be a few miles across or much larger. Fundamentally the model will try to predict how each quantity (temperature, velocity, humidity, etc.) will change over time in each box. These changes are averaged over the entire box and over the entire time step. The model starts with the current time, then adds the changes to estimate the value at the next time. Then it does the whole process again. And again. Within each box, each quantity (say temperature) can change because it's being added to or subtracted from through the walls of the box, or from internal sources or sinks.

Models have to think about the sizes of the time steps and each grid cell, and the overall size of the domain. The results are very sensitive to how boundary conditions and initial data are entered. There are always things that occur at scales too small to be resolved by the grid, so they have to be approximated. Turbulent mixing—like wind gusts—is one of these. In the atmosphere, you need auxiliary models to figure out if clouds, raindrops, or snow will form, and how those will interact with the overall system.

Greg went on to explain how modern weather models are *dynamic*, using input from live observations to correct themselves in midstream. This did not begin to happen operationally until the 1990s, when computers got big enough to handle the additional calculations. Talking physics with Greg feels a bit like standing across the net from Serena Williams, but his point about adding live observations rang with instant familiarity. That was me. Like most commercial mariners, I've recorded what feels like a lifetime of weather values, tabulated for transmission back to atmospheric agencies through the global Voluntary Observing Ship Program.

These precious data, I am assured, are piped into the regular run of forecast modeling, where they help to keep the equations honest. Digital devices have streamlined this process somewhat, but the ritual of weather reporting is still typified for me by the image of the rain-soaked sailor clutching a grease pencil and squinting to read the barometer with a dim flashlight.

The European Centre for Medium-Range Weather Forecasts (ECMWF) has its own model, roughly equal to the GFS but different in a few key aspects: ECMWF runs just twice daily, meaning that its inaccuracies may take longer to sort themselves out between runs. On the other hand, the EU model uses a finer data grid and a more sophisticated method for modeling pressure and altitude, giving it what many say is better resolution and long-range accuracy. Joe Sienkiewicz credits European leadership for significant refinements in dynamic modeling methods. And for now, they have a bigger computer. Some will tell you that the ECMWF is correct more often, which is not to say that it is never wrong.

The expected accuracy of a forecast drops off rather sharply with time. This doesn't mean that a model is wrong when it fails to get Friday's weather right on a Monday, only that the uncertainty of the system is too great to predict its fate without fail. As an end user of weather forecasts, I am generally hesitant to make firm plans around predictions that reach more than seventy-two hours into the future. The Ocean Prediction Center publishes its daily maps out to ninety-six hours, a point at which the prudent mariner understands that the real ocean may differ significantly from the predicted one. Statistical graphics from NOAA's Environmental Modeling Center bear this out, showing how the reliability of common models—what forecasters call *skill*—decays with each twenty-four-hour interval. Joe suggests that for now it's realistic to say that individual systems can be roughly forecast to about five days, and larger patterns to a week. NOAA distributes certain products that are shown as valid out to eight days, but no doubt does so with some hesitation. The GFS model itself can be run sixteen days into the future, yielding predictions that might offer some guidance but are not fit for operational decisions. Do not stake your life on them, or your livelihood:

"Should I plant my garden over school vacation?"

"Maybe. It looks like that might be a rainy week."

"Should we commit to a backcountry ski traverse of Stevens Pass two weekends from now?"

"Forget it."

Edward Lorenz was an American meteorologist, famous for his work in the development of so-called *chaos theory* and for naming the concept of the "butterfly effect"—whereby a minor disturbance in a complex system cascades into a range of unpredictable outcomes. He started down this path somewhat accidentally in 1961, when in the process of restarting a weather model on his computer he chose to round off a data figure from 0.506127 to 0.506. The model ran away in an utterly unexpected direction, leading Lorenz to his conclusions about just how sensitive atmospheric simulations were to small variations in their starting conditions. In this devoutly nonlinear world, he determined, the chance of forecasting with any skill at all would drop to near zero at about the two-week mark.

"With the chosen values . . . ," said Lorenz in his primary paper, "cumulus-scale motions can be predicted (up to) about one hour, synoptic-scale motions a few days, and the largest scale a few weeks."

The explosion in computing power has allowed models to gain about one day of predictive capacity each decade since the 1980s, but it appears that there is indeed a limit very close to Lorenz's threshold, a point at which the number of possible outcomes becomes too large to yield a stable solution. Forecasters use so-called *ensemble* methods as a hedge against the butterfly effect, running multiple iterations of a scenario in parallel, each with small variations on the initial conditions. This creates an array of possible futures, a set of multiple forecasts that are then averaged into one. It's a useful technique if you're trying to identify areas under threat from a discrete weather event, like a hurricane. This branch of statistics is called *Monte Carlo analysis*, a not wholly re-assuring moniker to those of us playing with high stakes at the weather table.

NAVIGATIONAL channels are marked with buoys, guiding objects set in place by careful people with precise instruments. Nonetheless, mariners are admonished not to place all their trust in a single piece of information: Follow the buoys, but also watch the radar, check your GPS, and look out the window in case some unknown event has left a marker out of position. Sailors seek this same level of redundancy with regard to the weather—recognizing that forecasts may be derived from a process named after a famous casino and that multiple outcomes are thus possible. If competing models predict the arrival of a storm at two different locations on the same day, it might be wise to avoid both until the future is better resolved.

There are some challenges baked into this cake. One is that the original authorship of forecast material is often obscured by layers of media. Many agencies, public or otherwise, obtain their primary data from the same basic suite of models. This means that in comparing a NOAA map to something you've just down-loaded from Weather.com, you could effectively be looking at the same thing twice. To achieve real redundancy, the imagined mariner (or airplane pilot, or backcountry traveler) must look in an informed way for alternatives. Most of all, they should have a look at the sky to see if what's there matches up with what the experts are predicting. Like so many things delivered by the internet, weather maps are beautifully rendered, packed with information, and rapidly obsolete. Think about this: When a model is used to draw a twenty-four-hour forecast, it is working with data that might be six hours old when the map is published. By the time the map itself is twelve hours old, it is a prediction of the future based on estimations made well in the past. This is OK, as long as you recognize that what you are seeing is unlikely to be the exact course of events. If this caveat is disregarded, the results can be costly.

A tragic example of such an oversight exists in the 2015 sinking of the American merchant ship *El Faro*. *El Faro* sailed on September 29 from Jacksonville, Florida, on her weekly run to Puerto Rico—planning a route east of the Bahamas that by all measures would involve a very close shave with a developing tropical cyclone called Joaquin. Less than forty-eight hours later, Joaquin had exploded into a category 4 hurricane, and *El Faro*—after failing to alter her voyage plan accordingly—foundered and went down with all thirty-three hands. It would be the worst disaster involving an American vessel in more than half a century.

The story of *El Faro* is a case study in poor decision-making, but the nature of weather forecasting in the digital age played its own part in the awful outcome. To start with, the beginnings of Joaquin had been brewing in the North Atlantic for some time, and the predictions for its future were far from conclusive. There were simply too many unknowns connected to other forces nearby in the atmosphere. In fact, the models were having such difficulty nailing down a forecast for Joaquin that the NOAA storm discussion on the morning that *El Faro* sailed ended with this unreassuring sentence: "Needless to say, confidence in the details of the track forecast, especially beyond 48 hours, is extremely low."

To mirror this uncertainty, the maps showed Joaquin milling about in more or less the same place off the southeastern Bahamas for days to come. At a glance, such a picture can make a storm look like a stationary object, but in some cases what's really being expressed is serious doubt about its future. Reverting to the old metric of the mariner's 1-2-3 rule—where each passing day introduces sixty miles of uncertainty to a hurricane's forecast position—there was by the first of October an equal likelihood that the center of Joaquin would be located anywhere within a circle close to three hundred miles across.

Captain Michael Davidson took a look at this information and made a plan to sprint southeast, past Joaquin's inshore flank on his way to Puerto Rico. This at best seemed like the running of a risky gauntlet, but *El Faro* was a fast ship, and there appeared to be just room for her to slip between the storm's meandering track and the hard edges of the outer Bahamas. Regrettably, Joaquin was in the process of drifting farther to the south and west than anyone had predicted—and as it moved into an area of warmer water and less wind shear, the resultant phase of intensification would double the storm's embedded wind speeds to 140 knots in less than forty-eight hours.

El Faro's hull was old, but her communication gear was modern, and the captain was planning his voyage with forecasts prepared by a commercial routing service and downloaded from a satellite email connection. What he was not aware of was something that came to be called "a lethal latency of information" in the hearings that followed the loss of his vessel. The data being sent four times daily to Captain Davidson's inbox had been collected from the model runs, bundled into graphics, and compiled in a way that meant some were at least six hours old—one complete forecast cycle—by the time he saw them. Once he was inadvertently sent the same map twice in successive transmissions.

The crew—watching the TV Weather Channel in their cabins and reading text forecasts on the bridge's INMARSAT-C receiver— were getting newer and more frightening information about the developing storm. Doubts were voiced, but in what was characterized as "a deadly failure in *El Faro*'s culture of communication," the crew's concerns failed to register in the captain's plan for action. Amid all this, El Faro's officers had no means to measure the strength and direction of the wind in real time, and thus no effective way to see that Joaquin was in an unexpected location and far more powerful than anticipated. The ship's anemometer was broken and had been so for months.

A T the end of the day Joe and I walk outside onto the center's broad surrounding plaza, a captain and a weather wizard suddenly just two commuters in the teeming diurnal tide of suburban Washington. It is May, the trees already a deep summer green and the sky boiling with clouds that would alarm me if I were at sea. Raindrops are just starting to dot the dry concrete in dark splotches. Joe heads for his car as I turn the hood of my jacket up optimistically and depart into the maze of sidewalks and cross-streets that brought me here. A navigator by trade, I am on land still sometimes deserted by my compass. South of me somewhere is the Metro station, surely—but my phone is out of juice. I have no chartroom to visit, no swarm of seabirds flying helpfully in the right direction.

I eventually find an intersection that I remember and set my course as the cars hiss by and a gust of wind makes its way through the treetops. There in the distance is my harbor, the glow of a Starbucks under the concrete bulk of a parking garage. A thick sheet of rain meets me halfway there, so I am soaked through by the time I find refuge. I line up for coffee beside a neatly dressed woman who is mysteriously dry, a picture of professional composure. Does she have a force field? I wonder. No, it is an umbrella—now folded smartly in the straps of her handbag. In fact, most of the people around me have umbrellas and are similarly squared away. Maybe they have read the weather forecast. Suddenly I am conscious of my wet footprints and streaming orange raincoat. This is the same jacket I wear at sea, the pockets still filled with old bits of twine and candy wrappers. It feels like I've walked through some portal from a parallel world, from a land where it's OK to dress like this for work into one where it isn't. Nobody seems to notice.

7

VOLTA DO MAR

The brigantine *Corwith Cramer* is the Atlantic sister of the *Robert C. Seamans*. Built for a similar mission, she is older and a bit simpler in her equipment, her lines somehow more graceful, with the distinction that seniority often seems to bestow on built objects. Standing on her quarterdeck, I pause to watch another ship that's in the process of arriving, just across the harbor from us. Two hundred meters long, she spins in place and backs neatly into her berth like a bread van—thirty thousand tons of steel, brought deftly to rest by the art of shiphandling. There is little fanfare surrounding her completed voyage, just a metallic *boom* as doors open at the bow to disgorge an instant stream of vehicles and people. "Grimaldi Lines," say tall white letters on the bulky hull, painted there like script on a billboard. Heirs to the Phoenicians, the Grimaldis can haul goods and people by sea anywhere that highways meet the water's edge. In just a few hours the vessel I'm watching will be loaded once more and bound back whence it came. Rome, perhaps. Sardinia, Toulon, or Ibiza. This is "short sea shipping," the coastwise trade that eschews roads to connect cities by the direct path of water. It is an ancient business made modern by craft like this one: fast as cruise ships but able to open mouths like basking sharks and swallow their cargo in prodigious gulps, three hundred cars at a time. Such ships are the connective tissue of Europe's perimeter, cheaper than air transport and often faster than road.

There is a gust of wind and a matching puff of smoke from the Grimaldi ferry as her master engages power to assist with the final steps of mooring. Rain drives in under our awning, and I step back to stay dry. It's early morning and there is no traffic on our own dock, save a sanitation worker in a bright jumpsuit sweeping bits of debris from around park benches. She makes her way past and regards me with a serious expression, a tall woman with a spectacular mane of dark hair and gold hoops in her ears. This is after all Barcelona.

In the deckhouse raindrops are coursing down the portholes in miniature rivers, and the floor is damp with footprints. The science team are there having coffee, looking idly over charts. Rosa Balbín nods hello to me, her small silhouette blurred in an old sweatshirt. She is an oceanographer from the paradisiacal island of Mallorca who is joining the ship here as an observer, a walking treasury of local knowledge. I smile good morning as the breakfast bell rings somewhere below, and I hear the engine room door open and close in response—two rattles of hatch gear with a burst of machinery noise sandwiched in between.

We are embarking on this voyage from a place where the interplay of humans and water runs back as far as one would care to look. Barcelona began as a Paleolithic fishing camp and is today a metropolis, a shining monument to marine commerce and reinforced concrete. Out beyond the cathedrals and Gothic alleyways of town is the container port, its paved expanse growing exponentially to keep time with the appetites of twenty-first-century

Europe. Two million containers have passed through here in the last twelve months, I learn, equal to a line of trucks stretched from Paris to Vladivostok.

Later we go walking along the esplanade at Barceloneta, where the crowds have vanished and waiters in their cafés roll plastic curtains down against a fresh tempest of drizzle. A windsurfer is racing back and forth parallel to the beach, his one upright sail like the wing of a dragonfly. Today's wind is a *llevantade*, Rosa tells me, a type of fall storm that develops when warm dry air migrates north from Africa and pulls moisture up from the sea surface. Buoyantly unstable, the air rises, cools, and releases water as clouds and rain. A loose nucleus of low pressure develops, set spinning by the Coriolis effect. Typically these systems park somewhere in the space between Spain and Sardinia, delivering a wet and boisterous stream of wind from the east, unpleasant enough to make you reconsider your plans, whatever they are. Rosa calls the Mediterranean "el Mer en una Botella"—the Ocean in a Bottle—a little inland sea with its own version of nearly every marine weather process, captured by plate tectonics and amplified by the variety of surrounding landscapes.

I hear something similar to this from someone else I meet not long afterwards, a bulky Croatian with a biker's mustache and arms like cordwood. He and a crew of equally imposing characters are en route to South America with a new tugboat, but plan be home again in time for boar-hunting season. Captain Marko's real passion turns out to be sailing, however, and we spend an hour sharing sea stories—mostly about his time spent cruising along the Balkan coast. It's a spectacular place, he tells me, dotted with islands and beautiful seaside towns. The Yugoslav wars were a setback, to be sure, but the only thing to be truly careful of now is the occasional *bora* wind. The bora occurs when cold dense air accumulates over the Eurasian steppes and funnels out onto the Adriatic Sea through the passes of the Dinaric Alps. It's a simple case of pressures and restricted flow, like a nozzle. Simple or not, the bora is ferocious, and can blow with little warning at speeds up to 90 knots. Even for captains on holiday, recreational sailing is not recreation in such conditions.

The bora is a Slavic cousin to the *mistral*, which roars down on Spain from France when high pressure builds over the continent and a low develops over the warm Mediterranean. The outflow funnels through the Rhône valley and ejects across the Gulf of Lyon in a narrow band, sometimes at hurricane strength— enough to threaten even large modern shipping. Mistrals reach as far south as the Balearic Islands, where they are locally known as *tramontanas*. I will be happy not to meet the mistral in person if I can manage it. It is just now October—a bit early in the season for concern, but not impossible.

Taxonomically the bora and mistral are *katabatic* (downhill) winds, found anywhere that cold mountain air can make a steep escape to ground. I'd heard of such things before in other places. Long before I sailed her on my own eventful trip to the Arctic, the schooner *Bowdoin* spent World War II in Greenland doing surveys of a long fjord called Kangerlussuaq for the US Navy. Her captain, a young officer named Stuart Hotchkiss, had been warned about something called the *foehn* wind, produced by downdrafts from the high frozen interior. Air swept in by polar easterlies over the ice cap would compress and actually warm as it forced its way down the valleys at great speed. The first harbinger of a foehn event was a sudden increase in temperature at sea level, accompanied by a rising barometer. This is exactly what Lieutenant Hotchkiss and his crew saw one fall afternoon in 1943, and they did their best to get ready, stowing loose gear and sending extra lines ashore. Sure enough, wind came as advertised within the hour, blowing 80 knots and threatening to beat their little boat to bits against the wharf. The first sign of real trouble came when the fasteners—long iron spikes and hardwood pegs called *trunnels*—began to work their way out of the hull in response to the punishment. All the men could think of was to pound them back in, and that is what they did, taking turns leaning over the side with a sledgehammer each time their tortured craft rebounded off the pier. "All good fun," Stu told me over lunch one day, armed perhaps with the equanimity that the passage of time can permit.

Wind names in the Mediterranean derive largely from geography. *Llevantade* has roots in the Spanish verb *llevar* (to rise) and is one in a family of winds that originate from the east. Rosa names the winds from elsewhere on the compass, familiar monikers that remind me just how much is going on in this crowded junction of continents and ocean. *Poniente* means west in Spanish and denotes fair breezes that blow in off the Atlantic, funneling through the Strait of Gibraltar. The *sirocco* is drawn up from Africa, a gritty inhalation that grows wet and foggy on a diet of evaporated water as it makes its way north. Microparticles of airborne sand form nuclei for condensation, bringing tiny bits of the Sahara down with the rain onto Europe. The sirocco is called the *arifi* (thirsty) in Libya, and the *jugo* (south) in Croatia.

In a few days our llevantade has gone, leaving just a lumpy swell that meets us as we idle past the breakwaters at the harbor entrance. Our students, newly joined, try mightily to walk in straight lines as the crew directs them from task to unfamiliar task. In midday stillness the sun is instantly bright, and other ships float around us like balloons in a smoky haze. There is no wind at all for sailing, the water oleaginous.

"Welcome to the Med," says Rosa. "One day of gales, nine days of calm."

I posit that we are experiencing the Mediterranean's unnamed breeze, the nonwind.

"Ah, yes," she replies. "El sin viento."

In our dining saloon the tables pivot on gimbals to counteract the motion, a traditional system that is effective but at first glance disconcerting. The ship rolls to port and the tables tilt to starboard. Mashed potatoes, salad, gravy, and lemonade all stay in their places as long as nobody touches the table. For someone in the early stages of seasickness it is an uncomfortable thing to observe. Attendance at lunch is noticeably low, and some who are there sit down with the smallest of portions. They eat quickly and silently, excuse themselves, and move rapidly toward ladders leading to fresh air. Medicine asserts that seasickness is simply the product of disparity between the eye and inner ear, but for those who succumb it represents an unnavigable intrusion of despair: vertigo, exhaustion, an acute disinterest in anything but sleeping or vomiting. Like altitude sickness, it is an interval of adjustment, one with an end if you have the time to get there. After a few days most people are fine. Some, unpredictably, will never suffer at all. The otherwise unimposing individual from a landlocked state goes happily about their business, a sudden alpha amid suffering shipmates. There are drugs you can take, other remedies of which I am less trusting: candied ginger, chamomile, wristbands that claim magic feats of acupressure. I am usually fine until the odd point when, suddenly, I'm not—often on some benign day far into a trip when, worn down or just dehydrated, I realize that lunch is no longer interesting. Unlike the green novices, I carry the expert's gift of knowing it will end.

The sin viento persists for most of the next two days, as the vast built landscape of Barcelona fades to a dim evening glow and disappears. To the south are Tarragona and the Ebro River, its broad delta marked in the sky by a tall heap of clouds. A solitary oil platform burns its gas candle off into the night. I think of the

Roman galleys we saw in the dim brick sheds of a Barcelona museum—long open boats, with sails for when the wind was favorable and banks of oars for the frequent occasions when it was not. In the abrupt and deadly winters, it was considered prudent practice to haul your galley up onto the beach and fill it with rocks so it wouldn't blow away before spring. We drift past the tall, tilted wedge of Mallorca, its carbonate massif left from the last long-ago collision between Europe and Africa. There are rare bits of primeval Mediterranean forest still left here, and ancient olive orchards terraced into near-vertical hillsides with artisanal arrangements of stone. It is a World Heritage Site.

"Who picks these olives?" I once asked.

"Nobody, anymore," I was told. "It is too much work."

In afternoon great baulks of cumulonimbus blossom above the island, drifting over us as rain. Ribbons of lightning cross the sky. This is *orographic lifting*, where wind-driven air is forced upward by interaction with a land mass. Pressure drops; the air expands, cools, and releases water as condensation. When the *Corwith Cramer* returns to America two months from now, the crew's first view of the place will not be land but clouds—great piles of wet tropical air pushed west by the trade winds and shunted aloft by the abrupt volcanic slopes of the Antilles.

In my cabin after dinner I am deep in some captainly task when someone comes to get me.

"A ship is calling us on the radio."

I go up to look. It is early evening, the sun faded to a dim orange band in the west, giving way to gray overhead. The extravagant lights of a cruise ship are gliding by to starboard, and sure enough, the radio is crackling:

"Sailing vessel *Corwith Cramer*, this is the cruise ship *Mein Schiff 3*, 9HA4883."

The mate reaches to answer.

"Yes, this is the *Corwith Cramer*, WTF3319. Shift to channel 10 and go ahead, please."

"Good evening, *Corwith Cramer*. Is Elliot Rappaport aboard?"

The mate looks at me like I am James Bond and hands me the microphone. I have no idea.

"Yup, this is Elliot."

"Elliot! This is Todd. What's going on?"

It is an old shipmate now on a much different career path than my own—at his console high atop a gleaming conveyance where five thousand people browse buffets, shop for fashions, and wheel service carts from cabin to cabin as he speaks.

"Todd! What a pleasant surprise. I had no idea you were doing this now. It looks like you've found yourself a good job. No more standing watch out in the weather, anyway."

"Well, yes, it has its advantages. I do miss sailing . . ."

We sign off and I go back below. It has started to drizzle, and as I duck through the hatch, I think about what else I might be doing with my life, trying to envision Todd's world as an alternative. I imagine better coffee and a spacious private cabin, closets filled with spotless white uniforms that someone else has washed.

❧

As we round the southeastern corner of Spain a day later the wind comes up to meet us, stiff and in our faces. The Alboran Sea is the pylorus of the Mediterranean, a long horizontal slot bounded by the adjacent mountains of Spain and Morocco. Even a small barometric difference between our bottled ocean and the one outside is enough to create significant wind amid the steep topography. In summer the Bermuda-Azores High typically forces air in toward the slack pressures of the balmy Med, filling the bottle until periodic reversals drive things back the other way.

The outgoing levant winds are cool and damp and send a banner of cloud streaming downwind from the Rock of Gibraltar. Either phase of this cycle can persist for days, which means that timing is everything if you are a sailor trying to gain or leave the inland sea. To complicate things for us, a compact low-pressure system has developed today off Cartagena, spawning a localized westerly gale with a moniker of its own: *el vendaval*. The vendaval, says Rosa, is like the poniente but not nearly as pleasant. I agree.

The surface of the Mediterranean is lower than the Atlantic by as much as twenty centimeters, a declivity created by prevailing winds and the rapid evaporation of this warm salty lake between Europe and Africa. The effect is most noticeable at the Strait of Gibraltar, where surface currents run steadily eastward in a flow that peaks at each high tide, like the slow pulsing of blood in some great aorta. Combined with the vendaval, this is today making our navigational goal feel a bit like digging a tunnel with a spoon. A well-handled modern yacht might hope to sail upwind in such conditions, but for a large traditional ship like ours such hopes are unrealistic. Sailing ships, by and large, are built to sail on fair winds and wait out foul ones. Lacking time to wait, we fire up our main engine, using all the power we think prudent to drive ourselves up and over the freshly steep seas. It is a fight to gain ground an inch at a time, not at all relaxing. Would the Romans be rowing in this? I ask myself. Probably not. More than likely they'd be sailing home or headed for the beach. In our own galley, dishes fly.

After dark the handle of the Big Dipper stands upright above the mountains of the Costa del Sol, an escaping constellation from the galaxy of lights along the shore. Our diesel gets warm under the load, and in the engine room a faint sauna smell becomes apparent. The noise is deafening. A gauge protruding from the manifold like a meat thermometer tells us that the exhaust temperature is at just over 850 degrees Fahrenheit, enough to set a couch on fire. The engineer assures me that this is warm but not too warm—he has some recent and perhaps unfair experience to measure by, since the ship suffered comparable headwinds on its way into the Med eight weeks ago. This would have been a levanter, just the wind we would wish to have now upon leaving. The *Corwith Cramer* is in the midst of being twice unlucky, at least in this regard.

It is dawn. From Gibraltar's limestone redoubt long rolls of cloud spin down from the upper ridge of the rock, cascading over ships at anchor. The structures of town are almost invisible at our distance, a long low sprinkling of rooftops and facades in the hazy light. From elsewhere on the continent you can travel through Spain to this odd enclave, and after clearing a border post find yourself suddenly in England. On a main avenue of chip shops and souvenir kiosks the throng of cruise ship passengers pushes past in sandals, buying snacks and riding a tram up the rock to have their backpacks looted by wild monkeys. Hidden along quiet side streets are old synagogues and plain doorways marked by brass plaques, discreet nameplates that give no real clue as to the business of their owners. In dim bars men in leather jackets play slot machines and chat rapidly in Arabic over tiny cups of coffee.

Merchant ships steam back and forth through the strait, fifteen per hour in each direction—enormous vessels, indifferent on all but the worst days to the challenges we are experiencing. Half the world's sea traffic comes this way each year, a ceaseless line of prosaically named craft bound for every corner of the globe: The bulk carrier *Arklow Spray*, headed for Port Canaveral, Florida, to load cement. The container ship *Maersk Hartford*, for Algeciras. The liquified natural gas tanker *Sonangol Sambizanga*, destination Soyo, Angola. I learn all this from a plotter screen in our chartroom—where ships appear as icons in a colored parade, their secrets revealed at the click of a mouse. Today vessels at sea interrogate one another through something called AIS (for automatic identification system), another dully named miracle of the data age that joins VHF radio and GPS signals into a shared net of knowledge. From a glance at a screen, the mariner can learn all they need to know about nearby traffic, along with a good bit of relevant arcana: One query returns a ship's name, position, course, and speed. Also its flag of registry, destination, length, tonnage, and operational type. Technology has transformed the sea's anonymous expanse into something much less opaque—a ship in mid-ocean is still alone, but largely unable to hide. Navigators must still act appropriately and follow the rules of marine traffic, but along with radar the AIS makes it very hard to claim ignorance if you are involved in a collision.

Not so long ago, obtaining such information was the stuff of fantasy, and ships relied heavily on halting radio exchanges, often misunderstood:

> *This is the sailing vessel* Corwith Cramer, *WTF3319. We are steering 270, speed 4 knots, calling the eastbound merchant ship on our port bow, twelve miles south of Cabo de Gata. How do you copy?*
> *Hello?*
> *Good evening, Captain, this is the sailing school vessel* Corwith Cramer. *Please confirm you are eastbound at the Cabo de Gata Traffic Separation Scheme, your speed approximately 12 knots?*
> *Portside ship . . . hello? This is the . . .* Asphalt Princess, *3EKY3 . . . Come back, please?*
> Asphalt Princess, *good evening, this is the* Corwith Cramer. *We have you on our port bow and expect a closest approach of one mile in twenty minutes. Please confirm you intend a port-to-port passing with us.*

Yes, portside ship, Corwith Cramer, *hello?*

Yes, Asphalt Princess, *this is the* Corwith Cramer. *Go ahead,
please. What are your intentions?*

Yes, no, I am someone else, I have no intentions. This is the
Asphalt Princess. *Good night.*

~~~

TUNA jump clear of the water alongside us, leaping up into
squadrons of flying fish and seabirds. It is forty miles more
from Gibraltar to open sea, past the shallow banks of Cape Trafalgar, and to get there we must endure another night of slow
pounding with the engine at maximum RPM. The air grows
steadily cooler and damper, a long swell now lifting our bow in
slow undulations. Pushing the cork aside, we have escaped the
bottle and gained the full-sized ocean. The decks are soaked in
spray, crusted with a fine layer of salt grown oily from evaporation. At dawn a broad bank of cloud sweeps past with the flickering fire of sunrise underneath, huge ships passing us in the
half-light like spacecraft from other worlds. By lunch it is calm
and sunny, cool enough to think of fall, relaxing in a way utterly
unconnected to the day before.

Outside the strait the stream of traffic splits into branches—a
thick stem toward the English Channel, another bound south
along the African coast, and a third pointed offshore, across to
North America or the Panama Canal. These ships will in the expanse of open ocean make their own choices, but for now they are
kept in order by the separation lanes etched in neat magenta on
our chart. We are headed west, then south as we change course
toward the Canary Islands—effectively cutting across the eastbound lane of ships in order to do so. Like a car turning left, we
must yield to traffic, on a street where the oncoming vehicles need
a mile to stop.

The mate gives instructions to our helmsman and takes us calmly through, making small adjustments and calling other ships on the radio if any doubt exists as to their intentions. It is as though we are no more than a bus merging onto a roundabout. Willy is a maritime academy graduate, at twenty-six already several years older than another man named William Bligh was when he sailed as navigator of the HMS *Endeavour* under Captain James Cook. The marine profession awards responsibility at an early age. In his years since school Willy has worked as a shipwright and as the mate on a tugboat, hauling barges to western Alaska for a company called Crowley Maritime. The Western Alaska Fleet, according to the Crowley website, is "designed to undertake challenges unique to Alaska such as uncharted locations, shallow river navigation, and limited marine based infrastructure." Here the term *marine based infrastructure* refers to such luxuries as wharves, cargo cranes, and roads. The Alaska crews work in alternating six-hour shifts, around the clock through the nominal arctic summer—dealing with ice, bears, and awful weather; routinely running their barges aground to unload in an operation known as "fuel over the shore," where hoses are hauled up the beach to tank farms stranded like rusty monuments on the gravel. Everything you own smells like diesel after these trips, I am told, though the money is good enough to simply buy new clothes when you get home.

Willy lifts a cup of coffee and looks over to confirm his plan for the next ship crossing our track. A light wind develops from the north, at first a gentle exhalation and before long enough to ruffle the sea surface and raise the occasional crest of foam. We shut down our engine and set sail, exulting in the sudden silence. Somebody brings their bedding up to dry, white pillowcases and a neon sleeping bag that billows optimistically in the clean blue light.

M Y grandfather was an aviator in the Air Force, and his long patrol flights would sometimes divert to Bermuda when the weather in New England got just too nasty. He kept a bathing suit in his kit for such contingencies, a fact my grandmother never completely made peace with. I was fortunate to know him far into my own adulthood, but my memory holds mornings from this time a half century ago, a towering man in a flight suit singing songs to the buzz of his electric razor. I thought he was an astronaut.

Bermuda, through a bit of geological unfairness, is the Atlantic Ocean's only atoll, perched by itself about 650 miles offshore from Cape Hatteras, North Carolina. The weather there is famously pleasant—dry and warm and just breezy enough, most of the time. It's perfect for golf, sailing, or gardening, if you bring your watering can. Climatically, Bermuda is like an orphaned piece of San Diego County, cut adrift and floating in the mid-Atlantic. This is not a coincidence, as Bermuda and San Diego both lie firmly amid stable *subtropical anticyclones*—atmospheric settling pools, where lifted tropical air is falling back to earth from high in the troposphere. The sinking air increases in pressure, growing drier as its remnant moisture is reevaporated. The result? Long strings of fine days, coupled with nights that cool as the ground gives back its heat.

This cascade of air pouring back to earth near the 30th parallel creates climate on a global scale in each hemisphere. Over continents it forms our great deserts—the Sahara, the Atacama, Death Valley—sere landscapes maintained by a steady downpour of dry air from above. Over the oceans, these falling columns spread like water poured out onto a floor. The Coriolis effect gives a spin to the diverging flow, steering the winds around in an ocean-sized loop: clockwise in the northern hemisphere, the opposite in the south. Bermuda thus sits often on a stable plateau of high pressure with a circular current of air at its rim. These are the so-called *horse latitudes*—winds can be light here at times, and getting becalmed in your sailing ship was hard on the animals. As an extreme measure, one legend has it, horses were sometimes jettisoned to conserve water.

The subtropical anticyclones are the anchors of air circulation in each ocean basin. The southern rim of the Bermuda-Azores High is the trade wind belt, a swath of tropical easterlies blowing steadily toward the West Indies from Africa. Approaching the Bahamas, the orbiting winds bend around to the southeast, and eventually the south and southwest, as they make their trip up the American side of the ocean. These are the dominant winds of America's Atlantic coast, steadiest in summer when the high is well established. After passing America, the winds recross the Atlantic as a swath of westerlies—at times quite strong—meandering along the polar front. Along the coasts of Spain and Morocco—at what would be the three o'clock position of the Bermuda-Azores High—are often northerly winds, enhanced by interaction with warm air parked over the land.

By evening the *Corwith Cramer* is flying along like a kite, under a clear sky with the wind astern. These are the *Portuguese trades*, northerlies that blow predictably in periods of settled weather, bending gradually eastward to become the true trade winds off the western bulge of Africa. The Portuguese explorers, astronauts of another time, used these winds and some educated guesswork to push European dominion out of the Mediterranean and into the world beyond. For the mahrineros of Lisbon, it was simple work on most days to sail south to places like Madeira and the Canary Islands, the first non-European stepping-stones of Iberian conquest. Getting home was harder, until someone took a gamble and found that if a sailor put his back to the land and sailed off far enough to the northwest, he might eventually make his way up into westerly winds and back to Portugal before the food ran out.

Known to sailors as the *volta do mar* (return from the sea), this discovery—rather like the splitting of the atom five centuries later—would have irreversible consequences for all that came afterwards. Christopher Columbus used an expanded version of the volta to get his fleet from Spain to America and home again, but credit for a bolder leap goes to Bartolomeu Dias, who tested the concept on a global scale. Charged with finding a new sea route to the East Indies, Dias sailed his ships successfully around the southern tip of Africa by betting correctly that the South Atlantic would have a wind system like the North did—including easterlies in the tropics and west winds in the higher latitudes. Dias's crew had seen enough by this point, but his discovery allowed Vasco da Gama—one of history's many brutal overachievers—to close the deal soon afterwards.

Da Gama did this in 1497 by sailing south from Portugal, across the doldrums and into the South Atlantic trades, proceeding on a southwesterly track nearly to Brazil before he found the west winds and headed back east for Africa. His fleet rounded the Cape of Good Hope and made its way into the Indian Ocean, where they found maritime commerce already well established between Africa and the Indian subcontinent. These ancient sailing routes relied on the Asian monsoon, a seasonal cycle of winds brewed by the annual heating and cooling of Earth's largest land mass. This repeating pattern let merchants ply sailing routes that were reliable from year to year, provided one had adequate local knowledge. Captain Vasco lacked this knowledge, but he had guns and might in fact have solved the problem by kidnapping a navigator and pressing him into service. Means aside, da Gama succeeded in reaching India, and over several subsequent voyages established a colonial toehold that would persist for 450 years—all the way until 1961, when the newly independent Indian state ejected the Portuguese from their last enclave at Goa in a two-day shooting war, now forgotten to all but the combatants.

Ferdinand Magellan's circumnavigation between 1519 and 1522 drew the final pages into the book of global wind, proving that the Pacific Ocean had a circulatory system much like that of the Atlantic. Magellan and his ships began their voyage by following a rough version of da Gama's route to South America. Working farther south, they beat their way against cold and stormy west winds, through the Patagonian archipelago, and out into the Pacific. Once clear of the land in this new ocean, the fleet sailed north past present-day Chile and Peru until they again found warm easterly trades. Extrapolating from his experience in the Atlantic, Magellan was confident that these winds would be there to find. Not all of his shipmates agreed, but those who didn't mutiny enjoyed a storm-free (if nearly endless) passage to the Philippines as they traversed the planet's widest body of water. The price was nonetheless high. Magellan's fleet was down to a single ship when it returned to Spain, with only eighteen of the 240 original crew still aboard. The commander himself was not among them, having been killed in a skirmish with some locals at the tail end of the Pacific crossing. His lieutenant Juan Sebastián Elcano was left to finish the job, sailing home with a skeleton crew, across the Indian Ocean monsoons and back up the African side of the South Atlantic. How simple that sounds now, put here on the page.

⁓

IN 1513 the Spaniard Juan Ponce de León led his first voyage to the place that he would eventually name Florida. Along with his legendary and perhaps apocryphal quest for the Fountain of Youth, Ponce de León sought to expand Spanish colonial interests and build his own fortunes beyond the reach of the Columbus family, who had taken a tight grip on the newly minted West Indies. No friend of the natives, Ponce de Léon would be killed by a poison arrow in 1521—not far from present-day Fort Myers and with his fortunes still in the balance—but by then he'd learned much about Florida, with its bizarre wildlife, crystalline springs, and a ferocious current racing past the future Miami at the ungodly speed of 4 knots. A ship could only hope to gain against this torrent by standing far inshore and riding the eddies, across treacherous shoal waters and perilously within range of missiles launched by unfriendly residents. Ponce de León's torrent was the Gulf Stream, the strongest branch of a circuitous ocean river that traces a clockwise path around the perimeter of the North Atlantic. While no scientist, he would still surely have been interested to know that such currents are the joint product of wind and the Coriolis effect—working together in a way that is not intuitively obvious.

Like so many natural processes, ocean currents were a mystery until the Norwegians went to work on them in the late nineteenth century. Fridtjof Nansen was born near Oslo in 1861, a founding father of oceanography best known for his heroic adventures aboard the expeditionary ship *Fram*. A massive hardwood bathtub built specially for polar research, *Fram* spent three years frozen into the ice north of Siberia in a bid to reach the north pole, drifting in a long westward arc after entering the pack near the sprawling delta of the Lena River. Frustrated with their progress, Nansen and a partner left the ship halfway through its float for an expedited attempt to reach the pole by dogsled. They never got there, but nonetheless made it back in time for a reunion with their vessel as she regained Norway.

One of Nansen's observations on this odyssey was that icebergs tended to move at an oblique angle to the prevailing wind direction. In other words, a region of strong easterly winds would produce a current that went northwest. A Swedish graduate student named Vagn Ekman was tasked with investigating this phenomenon, and in 1905 he published a fascinating model to explain the transfer of energy between wind and sea. The *Ekman spiral*, as it came to be known, worked like this: Energy is passed from the wind to the sea surface by friction, and thence downward through successive layers of water until all of its force is dissipated. This wind-affected stratum of ocean is known as the *Ekman layer*, and it can be imagined as a stack of spinning plates or clutch discs. Energy is transmitted between the layers, and at each transference the Coriolis effect imparts a slight directional rotation, depending on the hemisphere—hence the forcing at each depth is deflected several degrees to the right (or left) of the one above it.

Ekman added up the velocities of all the rotated sublayers and found that by the time he finished his math at the bottom of the wind-mixed region, the net movement of water was at right angles to the surface breeze. An east wind in the northern hemisphere will therefore move water to the north. A north wind will move water to the west. This process is now known as *Ekman transport*, an indispensable key to understanding how the atmosphere pushes the ocean around. Ekman transport is responsible for the phenomenon of *upwelling*, a thing which happens when strong winds blow parallel to a steep coastline. I learned about this firsthand off San Francisco one summer, where I found things much colder than I expected in a place with the equivalent latitude of Virginia, or Greece. Approaching the Golden Gate under full sail in July, I zipped up my parka and thought it some kind of a cruel joke as the frigid winds swept us inshore. The cruel joke was upwelling. In coastal California, northerly winds push surface water to the west, away from land—and what rises up to replace it comes from a deep cold place. This bottom water is full of untapped nutrients, which is why this kind of continental boundary is often connected to rich and productive marine ecosystems. Overfishing has taken its toll, but California was historically a place to catch plenty of tuna and anchovies—rather like Peru or Namibia, both similar coasts at the eastern rim of ocean basins, with prevailing longshore winds and deep water close to the beach.

In a mid-ocean anticyclone like the Bermuda-Azores High, with its rotating perimeter of wind, Ekman transport pushes water steadily toward the center from all sides. Bermuda, at the middle of its eponymous high-pressure system, is in fact surrounded by a veritable hill of seawater. It takes satellites to see such things, but here in the space age it's possible to measure a height difference of nearly a meter between the center and the edges of the North Atlantic. Bermuda has not vanished underwater because this piled-up mound of ocean has a strong impulse to flow back downhill, at a rate in rough equilibrium with the water being forced inward.

Water flowing away from Bermuda's bulge is deflected by the Coriolis effect into a clockwise pattern at the perimeter of the Atlantic basin. The current wobbles west across the tropics and into the Caribbean until, superheated and confined by geography, it gets ejected at great velocity into the Straits of Florida. This is the Gulf Stream, the world's strongest ocean current, and it's something to see as it pounds into the prevailing winds on its way through what is now a major shipping thoroughfare. The resulting conditions can be truly awful, irregular polyhedral waves that would make Sindbad seasick. Out in midstream eastbound ships can get a 4-knot boost, while traffic headed west is forced to hug the reefs in search of slack water, much like Ponce de León did long ago. A sluiceway of warm ocean, the stream makes its way north and east, turning offshore near Cape Hatteras and stitching a sharp boundary below the Grand Banks before winding toward Europe. Eddies spin off to the north as the current grinds past the continental shelf, lenses of productivity that swarm with fish and bird life.

Great Britain is never really warm, but despite its high latitude it is never really cold either, thanks to the warming influence of the North Atlantic current system. Weakened but still traceable, the flow past Britain turns southeast, and then south, returning to the tropics to begin another circuit. Depending on where you are, it may have a different name—the North Equatorial Current, the North Atlantic Drift, the Portugal Current, the Canary Current—but all are subparts of the same system. Oceanographers describe this wind-ocean system with a collective moniker: the *North Atlantic subtropical gyre*. It is the dominant feature of air and water circulation in its hemisphere. In fact, as the Portuguese captains of discovery predicted, there are multiple gyre systems, reproduced in each hemisphere of the great ocean basins: the North Pacific gyre, the South Atlantic gyre, the South Pacific gyre. There is an Indian Ocean gyre too, but as Vasco da Gama found, it is heavily affected on a seasonal basis by another large-scale feature, the Asian monsoon. Ocean currents are strongest on the western edge of gyre systems, due to something called the *western boundary intensification effect*. This is a product of fluid dynamics and Earth's rotation, poorly understood until a Dr. Henry Stommel of the Woods Hole Oceanographic Institution derived a series of models to explain it in the mid-twentieth century. The intensification is most visible in the powerful Gulf Stream and its Pacific cousin, the Kuroshio Current, where in each case a large land mass also helps to funnel things along.

The *Corwith Cramer* and her predecessor, *Westward*, have traversed the North Atlantic regularly since 1973, stopping with the plodding consistency of research vessels to tow plankton nets for a half hour each noon and midnight. The catch is reliably interesting. There are always copepods, animals from a family of tiny crustaceans that would rule the Earth if population were the only criteria. Also jellies, water striders from the insect genus *Halobates*, and the bizarre larval children of swordfish, eels, and lobster. The middle of the ocean is a busy place. Something else that appears frequently is plastic, sometimes in big pieces but mostly in the form of decomposed microparticles. My colleague Kara Lavender has been a leader of research to determine the extent of plastic adrift in the oceans worldwide, and the data give a spectacular visualization of gyres in action: Plastic is everywhere—but the bulk of it is far away from the humans who are tossing it into the trash. Instead, it is all drifting in mid-ocean, driven inward by converging Ekman circulation.

Woods Hole scientists were far from the first to notice an accumulation of floating stuff at the center of the North Atlantic. The same Portuguese sailors who rode the gyre to discovery also found large mats of brownish seaweed floating in its midst, a weed that they named *sargaço* after a plant they'd seen in their wells at home. It turned out to be a community of pelagic brown algae species, seaweeds that are unusual in their ability to live without ever attaching to the bottom. Raked in by the gyre and basking in subtropical sunshine, the rafts of sargassum are floating oases where remarkable creatures flourish. For entertainment one can put blobs of captured weed in an aquarium and marvel at the show—tiny scurrying crabs, rubbery nudibranchs, and voracious sargassum fish, sublimely camouflaged and hard-wired to eat everything they see. The mats of weed get large enough to spot from the air, gathering in long windrows like grassy verges stretched from horizon to horizon. This said, the so-called Sargasso Sea holds a lot more water than weeds, and the salty stories of ships trapped amid swampy tendrils are just urban legends from another time.

⌇

THE winds are warm and steady from astern as we close on the Canary Islands, the ship advancing with a weightless motion. It is our last night underway and I am dozing in my bunk, awakened occasionally by sounds of the watch conducting their business—the economical language of steering and sailhandling interspersed with slow ramblings of trivial conversation, the tales of small adventure and jokes now six weeks in rehearsal. We're ahead of schedule and I've left instructions in my night orders to take in sail and wait in the approaches until daylight, noting from the AIS screen that a number of other vessels are already doing the same. The business of harbors operates twenty-four hours a day, but like most enterprises favors morning. From below I feel the motion change and hear the ordered tumult of sail being struck, the ship suddenly quieter as the rush of moving water ceases. There is the muted drone of our generator from the engine room, and nearby the rattle of an escaped pencil now rolling across the floor, a minor hazard that I mean to correct before drifting back asleep.

In an hour I'm awakened again as we start our arrival. The light of dawn is just breaking, and ahead of us the mass of the island is palpable—a great hump of exceptional darkness in the west, looming above the necklace of city lights below. Our watch officer is on the radio with a tugboat just behind us in the queue.

"Tug *Taurus*, this is the sailing school vessel *Corwith Cramer*, WTF3319."

"Yeah, *Corwith Cramer*, the tug *Taurus*, KYV118. Go ahead, buddy."

"Yes, tug *Taurus*, good morning, Captain. Would you like us to stand off and let you go in ahead of us?"

"Yeah, tug *Taurus* here. Thanks for the offer, Cap, but that's OK. I got a drill platform in tow here and need to get some things sorted out. Y'all go on ahead."

The Gulf Coast drawl, ubiquitous anywhere that oil is involved.

We start up the channel and with the dawn behind us watch as *Taurus* and her tow take shape astern, a dot on the radar morphing into a chunky tug dwarfed by the city-sized object floating behind her. The distinctive rig silhouette is fluted with new sunlight, a labyrinth of pipe and iron next to a broad helicopter deck and accommodation structures where another crew must also be having their morning coffee. Las Palmas is the largest port in the Canary Islands, its massive concrete breakwaters reaching out from land like the arms of a great cactus. Here are vessels from all scales of human seafaring: cruise ships, yachts, car ferries, container ships, and the rust-streaked trawlers that fill every Spanish port, testaments to their nation's rapacious appetite for seafood. Among all this is a convergent swarm of sailboats, gathering for their yearly leap to America. Like us, they are poised at the threshold of the trade winds for a tropical crossing of the Atlantic, down a track worn by generations of passage-makers. Here in mid-November we are in the sweet spot for such a venture, with hurricane season nearly past and northern waters turning inexorably toward winter.

At the outermost wharves is the armada of the petroleum industry, less like ships than floating industrial sites—self-propelled factories built to probe the ocean bottom in search of oil. A technology called *dynamic positioning* makes this possible, a system of GPS receivers and directional thrusters that permits something the size of a Walmart to drift within a tolerance of meters as its drill turns miles below. These are anti-sailing ships, their wizardry akin to the stuff that lets modern aircraft land with their controls untouched unless the pilots choose otherwise. I have friends on these vessels too. Paid like lawyers, they join their ships by helicopter and stand watch at glowing screens as drill crews push lengths of pipe toward the bottom, slung from a derrick that is higher than any ship's bridge. In their offices below are copying machines, swivel chairs, and houseplants.

The Canary Islands are a kind of Spanish Hawai'i—a paradise for visiting millions, built on mountains blasted up from the deep by violent forces long ago. It is a place of microclimates and sudden shifts in terrain. In a short bus ride from the golden beaches of Las Palmas one can visit Caldera de Bandama—a great brown hole blown into the earth, a half mile across and two hundred yards deep. Its floor is a pocked expanse of suspicious-looking cinder cones, piled evidence of old eruptions. Once, on the nearby island of La Palma, I walked to the high rim of another such caldera, where in clear stable air the silver domes of observatories stood perched like ornaments. Trapped by a temperature inversion, layers of cloud glided up the mountainside and poured down into the crater's abyss like cream filling a bowl.

The name of these islands, Canarias, descends along a tangled path from the Latin root *canis*. This is the archipelago of dogs, not small yellow birds. The birds are from here too but were named later. The relative accessibility of this place means that people have been arriving for some time. First were the Guanches, Neolithic immigrants from what is now Morocco. They were not great seafarers, as far as anyone can tell, but with their homeland just to windward were favored by proximity. Next came agents of every Mediterranean civilization with boats and ambition—the Levantines, Romans, and Carthaginians, and finally the navigators of Portugal and Spain, sailing their caravels out into an expanding world. In fair weather, the recipe works as it always has: Leave

Gibraltar, turn left, and with the wind on your quarter after five or six days there are the Canaries—between whose jagged peaks are fertile valleys where anything will grow.

In our passage we have sailed without design between unfortunate mileposts of Spanish history. It was here in the Canaries that General Francisco Franco first consolidated power in 1936, before making his way north to overthrow democracy on the mainland in a three-year civil conflict that ended with the fall of Barcelona in 1939—one tipping point among many in Europe's eventual fall toward all-out war.

Today's continental rivalries manifest more benignly in the signage of competing restaurants:

"Wir haben Kartoffeln!" says one.

"We have fish and chips!" proclaims another.

At a tapas bar I order a snack in what I think to be fair Spanish:

"Buenos días! Quisiera unas papas bravas y una caña de cerveza, por favor."

The waiter answers in English.

"A caña? That is a small beer, sir."

"Yes, thank you. I understand, that is what I'd like, please."

"OK. Nobody who comes here ever wants a small glass of beer."

⌒

ON my first visit to these islands many years ago I met a family aboard their sailboat, which, like my own, was preparing to leave soon for the West Indies.

"It is an easy trip," they told me with typical cruiser's insouciance. "Just like the old saying. You will sail south until the butter melts, and then a right turn."

That is to say, leave the Canaries and finish the broad sweep of the Portuguese trades into the deep tropics, where one can count on the full power of the real trade winds for sailing west. When we put to sea, this is exactly what we found—three weeks of spooling miles with a warm wind astern, daily soakings from squalls but none threatening enough to make a sea story. On New Year's Eve 1990 we raised the tony enclave of St. Barts above the horizon and anchored soon after in the roadstead. Staggering ashore to celebrate, we were a roughshod horde, pushing past men in jackets and women in gowns to buy drinks at the bar. I remember a crackly and expensive phone call to my soon-to-be ex-girlfriend, far away in medical school.

"So, you've sailed across the Atlantic and back again," she said. "That's great! How was it?"

Now the *Corwith Cramer* is preparing to embark on a similar passage with a new crew, first back to the Virgin Islands and in spring home to New England. My own hitch completed, I run through a turnover with the next captain. He is reading the engineer's report and wonders aloud if the agent will keep his promise to have our fuel delivered by tomorrow. I can't say. We are tied up to a dock far inside the harbor, our egress blocked by a ship that— we have also been promised—will be sailing soon. Ashore nearby is the Maremagnum, a neon-trimmed hulk of bars and shops with tennis courts on its roof. Imagine the upper half of a cruise ship sawn off and dropped in a parking lot. I feel my kinship with the Phoenicians slipping steadily away. After dinner we make our way back up to the chartroom and conduct an official signing over of command, one brief and formal sentence written in the log. The new captain picks up his pen.

"Jason Quilter," he writes, "relieves Elliot Rappaport as master."

We shake hands, and at this I am absolved. Whatever comes next is not my problem, at least in the realm of problems that the captain must address. My only significant challenges in the near future will be finding a cab to the airport at four a.m., and then sneaking my stash of Spanish ham past customs at JFK. There is relief but also a mildly discouraging vertigo, the vanishing of an acceleration that has been my personal source of gravity for the last eight weeks. As another captain once said to me, one minute you're the man . . . and a minute later you're just a guy.

8

SAFELY OUT TO SEA

On my next trip to Spain I am reminded that the world's great sailing routes are less paved highways than patterns of occasional convenience, spun from the overall chaos of the atmosphere.

Cádiz is a sleepy peninsular city just west of Gibraltar, a perfect natural harbor at what was for millennia the doorway of the known world. Today the place is a cheerful sand-colored maze of cathedral domes and square towers, its long beach crowded in summer with vacationing Spaniards. Forts along the city wall host open-air concerts, and the cafés trade briskly in every imaginable species of seafood, all served with those tiny glasses of beer that never have a chance to get warm. It is October 19, 2016—just before Trafalgar Day, a British observance named for Admiral Nelson's victory over the French and Spanish fleets in 1805. You don't hear much about this from the Cádiz locals, but it happened just down the coast and more or less ended the contest for naval advantage in the Napoleonic Wars. It was a brief but epic confrontation. In a radical departure from established tactics, Nelson swept down on his enemy from windward and attacked at right angles to the enemy flotilla, crashing through the line of battle rather than merging obliquely for a more sporting side-by-side exchange of cannonballs. His opponents were shattered, broken into disarray by a navigational move that any Boston driver would have recognized immediately. The admiral himself was killed amid the fray and got shipped home in a rum cask for a hero's funeral. Shortly after the fight, much of the victorious fleet was in turn mauled by a vicious fall storm, scattered into chaos by the elements while basking in the brief glow of human triumph. You don't hear much about this meteorological coda to Trafalgar from anyone at all, particularly not British naval enthusiasts.

On this day 211 years later it is a somnolent afternoon, the air hazy with moisture after what has been a run of perfect clear weather. Midway through another fall cruise from Barcelona to Grand Canary, we are turning back to sea after three days spent wandering happily in town—days in which I was nonetheless bothered by a sense that an opportunity for progress was being squandered. The ship, heavy with a fresh load of fuel, handles sluggishly leaving the dock. A fender breaks free, and I feel the brief dull impact of steel on concrete before a gap opens between us and the wharf. The mate glances instinctively over the side to regard the streak of chipped paint, says nothing. One never comments when the captain scratches the paint. Within an hour a new item will appear on her list of chores for our next port stop. "Touch up topsides," it will say. "Rig new fender."

I am never happy when things get damaged, even in a small way—but today what I'm preoccupied with are the weather maps, imprinted as they are with the unappealing beginnings of a storm just west of France. The Portuguese trades, so reliable in summer, turn capricious by fall. Here the warm Iberian breeze can give way to nasty little cyclones, eddies of cold air spun off by the polar front as it whips back and forth between fall and winter. These compact disturbances often form in tandem with sibling storms off North America and are well positioned to ruin your day if you are—like us—leaving Spain with a plan to sail south. The predicted path for this system concerns me. A small meander in the jet stream over the Azores is forecast to become a more pronounced wiggle—teaming up with a matching feature over America to form the overall impression of an agitated snake draped across the weather map. Cold dry air from high latitudes will mix briskly with warm air from the mid-Atlantic, and the general flow appears primed to carry the results right in our direction. It is a textbook recipe for trouble. Instead of running freely with a breeze astern, we can expect strong headwinds as we pass through the new cyclone's southeast quadrant—conditions made to fling drawers open and launch pots of soup off the galley stove.

I am not thrilled at this prospect. Before we can commit to our upcoming plans, we need answers to certain questions. How far south is the new weather system going to travel? How powerful is it likely to become, and how long will it persist? Finally, what is the best move for us? I decide to phone a friend. When I reach Joe Sienkiewicz, he is on vacation, but even in his distracted state manages to sound ominously impressed with the state of things. He is constrained from giving out actual guidance but is happy to share his general thoughts, which in this case are not reassuring. A long warm spell in North America has left behind a lot of energy, he says, enough to destabilize the atmosphere and set its present whipsaw into motion. The storm we are watching has a good chance of strengthening and tracking to the south, more or less the same way we are hoping to go. The official forecast tells a similar story—with two nearly identical systems predicted to intensify on opposite sides of the Atlantic. Mirror images of unpleasantness, their pattern on the map resembles a Rorschach inkblot. By all accounts we can expect gale-strength headwinds by our third day at sea, sometime in the afternoon on October 23.

These types of storms, Joe reminds me, may intensify rapidly and can make big waves in a hurry. They are worst in their off-shore quadrants, where their deep wells of low pressure back up against the Bermuda-Azores High to form the barometric equivalent of a cliff. The rotary motion of winds in a compact cyclone can drive waves in multiple directions, making a crosshatch pattern of extra-tall peaks and extra-deep troughs. This phenomenon created mayhem in the notorious Fastnet Race of 1979, when a rapidly moving summer storm swept through a yachting fleet south of Ireland, causing nineteen fatalities. A friend of mine was in a similar storm in 1990 near our current position, an experience he recounted later with an image I have not forgotten.

"A giant hole opened up in the ocean," he told me, "and the ship fell in."

He and the ship both survived, but it is not an experience I feel eager to share as I review the forecasts: twelve-meter seas, some say, along with a reminder that "wave height shown is the average height of the highest one-third of the waves." Even my basic grasp of statistics tells me that on a day of twelve-meter seas, you're quite likely to meet a twenty-meter wave once in a while. During my time at college, the average height of students was probably about five feet, seven inches—but the tallest person on campus was six eleven, at least half a foot taller than anyone else, and I saw him all the time.

"What do you think, Cap—not too bad, huh?"

It is Jeff, the chief oceanographer. I am a frequent guest at his home in Woods Hole, where he gets up at dawn to play soccer in a league in which many of the players are half his age. At sea he lives for data—most of all the netted samples that open a window into the hidden universe of plankton. He is looking over my shoulder at the chart table, where two trainees are already working laboriously through our first round of hourly weather observations. Jeff looks like Lance Armstrong, with a less deliberate haircut. He is a consummate shipmate, the sort of person who would share his last chocolate bar with you in a lifeboat. I try to sound as helpful as possible.

"Yeah, the first twenty-four hours look OK. After that, I'm not so sure."

Jeff takes a swig from his water bottle. Everybody on the ship carries these now, as though this were a triathlon, or perhaps just out of fear that they'll be stricken by dehydration with no one near enough to help. The first sign of real rough weather for me is always the hail of loose water bottles, coming from every direction like unguided missiles.

"We can go out and get some samples in the gulf, right?" he says. "The plankton team are raring to go."

The plankton team, I think, are the proxy for Jeff's own irrepressible energy. I offer up my most encouraging answer.

"We'll see."

The low skyline of Cádiz is dissolving into the murky distance behind us. I imagine people on the city wall with their ice cream cones, watching a small white sailing ship set out to sea. The mate comes to tell me we are passing through the outer roadstead now, the muted bulk of anchored ships ghosting by to one side, and then the other. The bell rings for lunch as our own routine rekindles around us. It's comforting how quickly this always happens after the brash interruption of time in port, but today I am less than comfortable. Are we right to be leaving? I'm not yet sure.

When to go and when to stay? The proper response to storms in voyage planning is not in any table of solutions. No mariner would willfully endanger their vessel, but what constitutes risk for one ship might be routine for another. Captains weigh the steady pressure of obligations against the known and estimated hazards of their world, using a calculus whose terms alter from one day to the next. In Joseph Conrad's novella *Typhoon*, a master drives his ship directly into a horrific cyclone through pure lack of imagination. Every instinct from long experience warns him that deadly weather is in the offing, but he lacks the strength of character to change course—a decision that would cost his company time, fuel, and money. If he sailed around the storm, how could he prove to the owners that it was ever there to begin with? The consequences of his choice are devastating. The ship survives but is battered into a worthless hulk—and the men aboard, chief among them Captain McWhirr, are changed forever.

When things go wrong for vessels at sea, the events are typically investigated afterwards by some official agency of the ship's flag state. Inquiries can be lengthy, diving deeply into the chain of events to identify causative factors. Hard questions get asked. The findings that follow are by design meant to offer guidance rather than simply blame, but that can be a difficult line to straddle—and in accidents involving weather, doubt will always be raised as to whether the ship should have been there to begin with.

In December of 2006, the sail training ship *Picton Castle* lost a crew member overboard in storm conditions during a voyage to the West Indies. With a relief captain in command, the ship had left Nova Scotia three days previously and was making her way south after the passage of a frontal system. While patrolling the deck on the night of December 7, a deckhand named Laura Gainey was swept over the side by a boarding sea. She was never recovered. The authorities afterwards cited several causal factors, but placed emphasis on the question of why the ship had gone to sea at all. The following language is taken directly from the findings of the Canadian Transportation Safety Board, and suggests in the tortured prose of official reports that the master's decision to sail had not—in the investigators' opinion—been a sound one:

> *Although it was reported that long-range forecasts were taken into consideration to determine when conditions would be favorable to sail—the forecasts clearly indicated storm conditions for the occurrence time and area . . . Mariners need to be aware that, when sailing in winter months, weather conditions can change rapidly and can be quite intense. Consequently, when planning voyages, longer-range forecasts should be taken into consideration.*
>
> *The master was also aware that further delays, in combination with the onset of winter at sea that would make the voyage more difficult, could potentially jeopardize the voyage to the Caribbean. Crew and trainees had already been selected and payments made for various legs of the voyage . . . Given the financial benefit associated with proceeding with the voyage, this increased pressure to sail.*

Pressure to sail. Deciding when a ship can safely put to sea is a process fraught with variables, and one that is unfailingly revisited after a casualty. Well-found vessels are set up to endure most of the weather that they are likely to see, most of the time. Different ships have different missions, and varied thresholds for risk and punishment. The worst weather may best be dealt with by simply staying out of the way—but slow-moving craft on long passages are rarely able to dodge everything that comes their way. Big, fast ships have more options, but most of them are on tight schedules with major economic stressors. All navigational decisions thus take on an element of calculated risk. The weather can always get worse than you expect it to, a dangerous thing if safety margins have already been reduced by an optimistic bet on conditions. This is what I try to think about when I get a feeling that things are nearing some boundary of comfort or safety. How much worse can things afford to get, I ask myself, before we have a real problem? Changing a plan can bring its own risks. Diverting a ship from its original route may introduce unforeseen pressures elsewhere, and this complicates the choosing process. Most ships reach their destinations safely and on time, but some regrettably do not—and nobody wants to be part of a sea story.

When the merchant vessel *El Faro* was lost off the Bahamas in Hurricane Joaquin, the hearings that followed took six weeks and sought a long list of answers—first among them why a forty-year-old ship was still being permitted to operate as she was, with only limited updates to her mechanical and safety systems. The elephant in the room remained the mystery of just what Captain Davidson was thinking, brushing intentionally past a tropical cyclone to follow his normal weekly route to Puerto Rico.

Most mariners who are honest with themselves understand at least part of the answer to this question. He thought he could do it. He was being well paid to run a busy ship on a tight schedule, and no doubt felt significant pressure to do his work without costly delays and diversions. There was speculation that pressing too hard on safety issues had already cost Captain Davidson a previous job. He had a mortgage. In hindsight, Davidson made exactly the wrong decision and in doing so paid with his own life and those of thirty-two others. Along the way he committed the unforgivable sin of failing to hear the concerns of his crew, who stood watch on the bridge as the flaws in his plan grew steadily more apparent. But as we are warned, hindsight is 20/20. Many of the threads in a deteriorating situation are invisible in real time without an uncommon level of insight from the participants.

Determining whether it's safe to proceed may be the hardest decision in a master's playbook. A ship is safe in harbor, the saying goes, but that is not what ships are made for. Some cases are clearcut, while in others the right answers may be available only in retrospect. I would never have done something that stupid, we think. I would never lead my team to the summit without an adequate supply of oxygen. It is melodrama to frame each decision against the backdrop of disaster, but none of the players featured in the case studies were planning on an accident when their days began.

⌒

SOME years ago I joined a ship in Key West. It was late February, and while Florida is popular as a winter destination, the wind can still roar there. That's exactly what it was doing when I arrived—blowing a near gale from the northeast, directly against the brisk flow of the Gulf Stream. This clash of wind and current creates uniquely awful conditions in the Straits of Florida, with no two waves alike. I have heard local sailors call them "condos"—big, square, and ugly, they pop up from nowhere and collapse without warning. In this case I was the new captain, perhaps feeling like I had things to prove. The crew were tired of town and raring to go—or, better said, the wind wasn't forecast to ease anytime soon, and I doubted our preparation would be helped by additional days spent visiting the bars on Duval Street. We were berthed at a local naval base annex, where the overtasked dockmaster seemed eager to see us off. When he made his daily call to ask about our estimated departure time, I felt a little like Captain Cook on his last visit to Hawai'i.

And so we went. Sadly, northeast was exactly the direction we needed to go—directly upwind. The strategy I chose was to take the ship out into the middle of the strait, where we'd get maximum help from the current in our battle to make progress against the howling wind. In two days, we went eighty miles. A roast turkey flew from the dining table into someone's bunk. The main engine overheated, and on the second afternoon a wave broke over the deckhouse and sent a sheet of green water cascading down into the chartroom. Miraculously our electronics survived. By the fourth day we were sailing happily toward the Bahamas across a smooth sea. People were drying out their clothes, and somebody was writing a song to commemorate the turkey episode.

The point of all this is to reflect on how fine the line can be between a serious problem and a good sea story. In my mental journal I rewrite this early cruise to the Bahamas, imagining a different outcome:

"American sailing ship suffers engine room fire off Fort Lauderdale," says the article in the *Miami Herald*. "Navigational instruments ruined by breaking waves.

"'I was tired of Key West,' says captain."

I like to think that I've become more comfortable with accepting the idea that it's sometimes a day to stay, not go. To my good fortune, I've worked mostly for organizations that are sympathetic to this approach—though the process of deferring old plans in favor of new ones can still be challenging, even without active resistance from the home office. We have learned from psychologists that our brains lay a kind of track when plans are made, a railbed that must first be uprooted before an alternate path is taken. The underpinnings of this tendency lie buried in our neural tissue, explaining in part why people run red lights, or even stick with bad marriages.

In the marine industry a practice called *bridge resource management* has been fostered to mitigate this known behavioral trait of humans. It's designed to convert hierarchies into teams, where everyone is briefed on the plan and—with proper training—feels empowered to intervene when the plan no longer seems to be working. There is something similar at work in operating rooms, where (we are promised) a nurse will tell the chief surgeon if they're about to replace the wrong kidney. Known for short simply as BRM, bridge resource management was developed in response to accidents like the loss of the *Herald of Free Enterprise*—a North Sea ferry that sailed from Belgium in 1987 with its front cargo doors still open, only to capsize within sight of the pier. Nobody on the crew had warned the captain of this error, and 157 people drowned. The BRM system has roots in commercial aviation, where some time ago it was decided that granting a captain unquestioned omnipotence was not sound cockpit practice. In 1977, two full 747s collided on a runway in the Canary Islands after one throttled up for takeoff and ran directly into the other. The captain of the first jet—KLM's senior pilot—was at the controls, and nobody thought to challenge him.

"If I start to leave the dock with our power cord still plugged in," I always tell the crew, "please let me know. And don't assume that I see that guy in the fishing boat so obviously crossing the channel right in front of us."

These protocols have helped to reduce accidents, but are not foolproof, interwoven as they are with other threads that affect human behavior—deference, complacency, fatigue, ego, and the pressure to perform. The guidelines of the International Maritime Organization call bridge resource management a "soft skill," dependent on the conscious participation of its members. It is a guide for conduct, not an engineered safety device. In his essay "The Ethnic Theory of Plane Crashes," author Malcolm Gladwell posits that rude Americans are far more prepared to call out an errant superior than might be the deferential Koreans or polite

Colombians. This said, bridge resource management did not ulti-mately save the *El Faro*, even as her crew stood their watches and wondered just what the captain was thinking.

⁓

L IKE her sister, the *Robert C. Seamans, Corwith Cramer* is a small but robust ship, built to perform in offshore conditions. She is well-equipped, with an experienced crew and a good record of dealing with all sorts of weather over her career. She is also punishingly slow, without the engine power or sailing efficiency to make much progress against strong headwinds. In addition to the professional staff, we are packed with college students—a precious cargo of fit and energetic young people who've had a lot of training and are ready for anything—or so they think.

By the afternoon of October 20 we are eighty miles west of the harbor, still motoring offshore in calm conditions. I gather the staff to talk about options. Based on the forecast maps, we have about forty-eight hours before the outer radius of the storm be-gins to affect our location. That gives us at least part of the next day to continue on our way. If the system cooperates and turns east toward the Spanish mainland, we'll by then have made enough progress to hold our position and let it pass inshore of us. And if it continues to push south as forecast, we can abort our passage and return to Cádiz with fair winds—arriving on the 22nd, well east of the worst weather. We stand on through the evening, slowing periodically for Jeff and his oceanographic team to collect samples. Early on the 21st we stop and wait, feeling like we are about as far from Cádiz as we want to be until a firm deci-sion can be made about proceeding. This step is not long in com-ing, as the new forecast arrives before breakfast. It says all I need to know—if not necessarily what I want to hear. The polar front is predicted to push even farther south in the next seventy-two hours, creating a near certainty that the surface cyclone will be headed exactly where we'd once been planning to sail.

Grateful for information that Vasco da Gama could not have dreamed of, I set a course back for Cádiz. We will wait for a better day, and not risk whatever consequences a needless beating might bring us. Jeff is disappointed but understanding. We arrive back at the harbor approach on the afternoon of October 22 in a sheet of driving rain. This is the warm front, its long thin wedge of tropical air reaching toward the center of the storm. There is a lot of energy in the air already, and the appearance of some waterspouts close astern leads to a bit of very efficient sailhandling when we are still about ten miles offshore. The visibility is rotten. I hear the harbor pilots telling someone on the radio that they are waiting for the rain to stop before they'll move a ship. Cádiz is not Seattle in its tolerance for precipitation. Gradually the city comes into broken view behind smeary curtains of cloud, and we make our way up the inlet under a new and spectacularly out-of-place suspension bridge. Our contingency berth is in a distant corner of the port, the agent waiting in his car under a dull line of empty grain silos, windshield wipers working against the downpour.

Few are crushed at the prospect of returning. For the next several days we drink Spanish coffee and watch the storm pass close by to the west, directly along the route we'd been hoping to take. It deepens as forecast, eventually separating entirely from the jet stream and drifting almost aimlessly over the eastern Atlantic. When we leave again four days later, the system is still blundering around like a zombie, now to the south of us and fully truncated from the main flow of the westerlies. Sometimes these cut-off lows can stay parked for days, making up for their diminishing intensity with sheer stubborn endurance. The wind is still near gale strength on our departure, though much more manageably from astern. Running off under shortened sail, we steer carefully, admiring the big leftover sea rolling in from the north. We adjust our passage plans for a direct transit to Grand Canary, a planned stop at the island of Madeira now a casualty of lost time. I call and let our agent at the port of Funchal know we won't be coming. With no small disappointment I imagine David on the dock under his Hermès umbrella, clutching a bottle of wine that we'll never get to drink. He is understanding to a fault and refuses any payment for the work he's already done on our behalf. It has been raining for days, he says, the steep alleys of town turned to whitewater creeks.

"You wouldn't want to be here now," he says, no doubt fearing for his home island's reputation as a verdant vacation paradise. "Come back when the weather is better."

9

A RIVER OF WIND

In August of 1947, the *Star Dust* fired up its engines in Buenos Aires for an afternoon flight to Santiago. The scene resembled something from a Graham Greene novel: a hulking piston-engined airliner thundering aloft in an exotic austral city while a small and mysterious cadre of passengers adjusted their seat belts in the cabin. Among them were a German widow with her husband's ashes, a Palestinian carrying a hidden diamond, and a British foreign service courier—a "King's Messenger"—on some opaque mission for the Crown. The air route from Buenos Aires to Santiago runs west for about seven hundred miles, crossing the South American coastal plain before hopping over the Andes to the Chilean capital. In terms of geography, a trip from Denver to San Francisco would be a rough northern analogue. The journey of the *Star Dust* passed that day under seemingly routine circumstances, and about four hours after takeoff the crew radioed air traffic control to report their imminent arrival at Santiago. That was the last anyone saw or heard of them until 1998, when a group of climbers came upon the plane's wreckage melting slowly out of an Andean glacier, some fifty miles east of her destination.

By then, the last flight of the *Star Dust* had taken its place among the great unsolved mysteries of air transport, the gamut of possible explanations running from Nazi sabotage to alien abduction. Using aviation science and a bit of forensic meteorology, the long-delayed investigation reached a more scientific conclusion: *Star Dust*, flying above the clouds at an altitude of 24,000 feet, had plowed into the teeth of an unanticipated headwind—a jet stream—and as a result the flight crew badly overestimated her progress. In all likelihood, they made a controlled descent into the ground, thinking their ship was well clear of the mountains when in fact she still had miles still to go.

Jet streams are high-altitude westerly winds, created by the sharp variations in density that exist in the atmosphere across different latitudes. They concentrate in narrow bands at the boundaries of the main global air masses and can blow hard enough to slow even twenty-first-century air travel. The most pronounced streams occur at the polar fronts, where temperate air in each hemisphere meets colder air from the Arctic and Antarctic regions. They are often portrayed in graphics as smoothly contiguous features, regular as roads, but the actual phenomena are really more like braided rivers or dotted lines—composites of multiple meandering subparts, adding together to produce movement.

At the time of the *Star Dust* casualty, an understanding of high-altitude winds was just starting to gel after several decades of observation by pilots and meteorologists. The first scholarly account of the topic is credited to an obscure researcher named Wasaburo Ooishi, who worked alone at a lab north of Tokyo in the early 1920s. Ooishi launched thousands of paper balloons and followed them through the eyepiece of his theodolite as the wind carried them up and away. From measurements of range and elevation he concluded that there were steady currents of westerly wind blowing briskly across the sky, high above Japan. They were most prevalent in winter, when they could be very brisk indeed—upwards of 100 knots.

A modern analysis of Ooishi's data has confirmed his findings, but at the time they were largely ignored by Western academia. He was a distant figure, far from the epicenters of contemporary science, and published his work in the suspect utopian language of Esperanto. Nobody in Europe or America paid him any mind. The first direct application of Ooishi's findings came in a bizarre campaign to attack the US with an armada of incendiary balloons launched from Japan during World War II. Ten thousand such devices were built in secret workshops and set adrift, equipped with ballasting mechanisms to keep them at the proper altitude along the way. At least three hundred of these strange weapons are known to have landed in North America during the final years of the war. One managed to kill six people when they stumbled across it during a church junket in the Cascade Mountains. Given that many likely fell unseen into backcountry, it's estimated that a thousand of the drifting drones, 10 percent of the total, probably made it across the Pacific on Ooishi's winds.

The jet stream wasn't given its actual name until 1939, when a German scientist named Heinrich Seilkopf coined the term in a textbook—but by then there was an established awareness that something was happening up there. As part of his work in the mid-nineteenth century, the American mathematician William Ferrel derived a set of equations that predicted the presence of strong winds aloft, based on data from surface measurements. Soon afterwards in Belgium an aerologist named Léon Teisserenc de Bort was able to verify some of Ferrel's theories using kites and balloons. By collecting temperature readings at altitude, Teisserenc de Bort also managed to establish the existence of the tropopause, a stable layer of minimum temperature about eight miles above the ground. He got himself into brief but serious trouble one day when a bunch of his kites, strung together on miles of piano wire, fell down in a snarl over Paris.

Wiley Post was a pioneering aviator of the early twentieth century, just the sort of visionary nutjob that Americans love to anoint as their heroes. Post missed the chance to fight as a pilot in World War I, but after brief diversions into oil field work and car theft, he found his way back into the sky. In the barnstorming '20s, he rose quickly to fame as an air racer and stunt pilot. He wore an eye patch and had spent time in jail. He also had a natural flair for engineering despite minimal schooling and correctly saw the future of aviation in the high-speed, high-altitude transport of mail and passengers. He was the first pilot to complete a solo flight around the world, and perhaps the first to wear a pressure suit. The latter made Wiley look like the Michelin Man in a diving helmet, but it permitted him to fly as high as fifty thousand feet during a series of transcontinental flight attempts in 1935. While none of these efforts ultimately succeeded, his instrument data led him to note that there were strong belts of westerly wind in parts of the upper troposphere. For this, Post is sometimes credited with discovering the jet stream, but it's fairer to say that he encountered it.

World War II pilots also encountered the jet stream, often unexpectedly. Much of aviation meteorology was at the time biased toward forecasting cloud cover (you can't bomb what you can't see), and while the concept of high-velocity wind aloft was not by then completely new, it was often absent from mission briefings. Jet streams are *zonal*, meaning that they flow mostly in a west-to-east direction. Allied aircraft returning from raids over Europe sometimes met extreme headwinds and were forced to ditch in the English Channel when their fuel ran low. American B-29s on the way to bomb Japan bucked winds of up to 140 knots. The streams could often be avoided with small alterations in course and altitude, but the navigational constraints of flying in formations made this difficult to do in practice.

The doomed *Star Dust* in fact vanished in the same year that the first comprehensive study of the jet stream finally went to press, published by a group in Chicago under the leadership of Carl-Gustaf Rossby. Rossby was a fascinating character, a central figure in the advancement of meteorology during the middle of the twentieth century. A brilliantly effective teacher and organizer—and reportedly a bit of a bon vivant—he loved long, lively restaurant dinners and was apparently never comfortable driving his own car. The Chicago group's first article, "On the General Circulation of the Atmosphere in Middle Latitudes," nailed a theoretical framework under the long database of high-altitude wind measurements. In the spirit of Rossby's collaborative approach, the authorship of this landmark piece was credited simply to "Staff Members of the Department of Meteorology of the University of Chicago," without the hierarchical listing of authors that is so common to scientific papers. Before diving into the facts, the document appealed for better communication among players in the new field of atmospheric studies, with a charming bit of old-world diplomacy:

> *There exists at present . . . a noticeable divergence of opinion with regard to the proper interpretation of several of the basic processes in the atmosphere. Because of this divergence . . . an effort was made to bring together research workers representing widely different points of view. Until a genuinely efficient method of data distribution to interested agencies has been set up, it is reasonably certain that Government funds invested in research institutions outside Washington will fail to yield maximum returns.*

Given that the second half of this statement could have been written in 2020 as easily as 1947, it's fair to say that Rossby's theories about the atmosphere have gained more traction than his advice on streamlining government-funded science.

⁓

THE troposphere—where most weather happens—could be imagined as a big room whose ceiling slopes downhill from the equator toward the poles, getting closer to the surface as the atmosphere underneath it grows colder and denser. This ceiling is the tropopause. It is not a smooth slope, but more like a series of gently tilted plateaus separated by cliffs. The steepest of these cliffs is at the abrupt thermal boundary between the cold polar zones and the much milder temperate regions. A second break sometimes shows up closer to the equator, at the border between the temperate regions and the tropics. In both cases, the sudden temperature change produces a steep difference in pressure aloft. Air flows across the tilting tropopause, and in the steep places it accelerates. The Coriolis effect deflects it eastward, creating a west-to-east channel of wind that is concentrated along the sharpest transitional zones of atmospheric temperature.

The Chicago paper focused mostly on the *polar* jet stream system, seen as the most powerful and influential. Rossby and his cohort described it as a distinct, surprisingly narrow belt of westerlies: "a meandering river winding its way eastward through relatively stagnant air masses to the north and south." They got especially interested in a shapely pattern of waves that formed curves in the circulation. Known typically as *long waves*, these meanders are today understood to be a catalyst for cyclones and a major mechanism for moving heat between latitudes. Rossby derived an equation to describe these features, and they now bear his name.

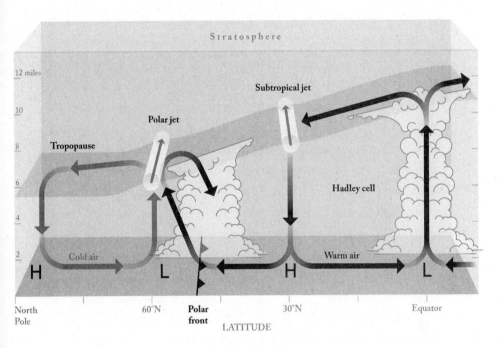

Fig. 9.1: How Jet Streams Form

A typical diagram of the atmosphere's lower layers, including the tropopause and the standard locations of the polar and subtropical jet streams. Drawings like this one greatly exaggerate the height of the atmosphere to get their message across. For example, recall that the troposphere is only between 4 and 12 miles thick, a tiny fraction of the distance between the poles and the equator.

When forecasters discuss the location of the jet stream, they are most often talking about the *Rossby waves*—which in their positioning are a primary determinant of weather on the ground. With their long meanders, the Rossby waves push cold air *troughs* into the temperate regions, while wave peaks—or *ridges*—drive wedges of warm air back toward the poles. Rossby waves travel eastward very slowly, but within the large-scale flow are smaller, faster-moving *short waves*, which Joe Sienkiewicz compares to ripples riding on an ocean swell. These shorter pulses of energy are the day-to-day triggers for surface weather. Bursts of cold air push toward the equator and sink to the surface as warm air is raised aloft, spawning cyclones by the process that the Norwegians called *frontal lifting* in their models. Surface storms are thus the eddies bordering a more powerful river of wind aloft, following a track more or less in step with the air streaming by above them. They are transient features, but if the long wave patterns behind them persist, they can happen again and again.

In the winter of 2015, a great cold trough parked itself like a frigid cow's tongue over North America, hatching blizzards one after another as subsequent pulses of warm southern air bloomed into cyclones. The cow's tongue moved on eventually, but not until most of the country had logged record amounts of snow. A shorter but equally brutal batch of cold air arrived in January of 2019, when Chicago was briefly the coldest place on Earth. Temperatures fell to near minus 50 degrees Fahrenheit and media outlets spoke excitedly of an assault by a *polar vortex*—a term borrowed in recent years to villainize especially unpleasant winter events over North America. In fact, the real polar vortex is a bitterly cold blob of air that resides high in the stratosphere each winter, typically ignored by the populace until like some mad bull it breaks free from its enclosure to wreak havoc in the towns. The Chicago freeze of 2019 actually began when warm air from Asia found its way up into the stratosphere and forced the normally stolid polar vortex to dissociate into lobes—one of which wobbled down over North America, taking a deep cold meander of the polar jet stream along with it. Called *arctic outbreaks*, these displacements appear to have become more common in the last half century, for reasons that researchers are still working to resolve.

MARINERS use the location of the polar front to estimate the extent of winds at the surface, making it an important marker in ship routing. The graphic of choice for this process is something called the *500-millibar chart*, which shows the changing height of the troposphere through a set of elevation contours, much like a mountaineer's map. On the page upcoming, two forecast maps from our aborted Cádiz departure are an excellent example of how jet streams and surface weather are connected. Figure 9.2, a 500-millibar chart, shows deep troughs developing in the polar front on both sides of the Atlantic. Coupled with a large blob of warm air that has settled over mid-ocean, these mirrored features are like paired eddies—spots where cold polar air is bulging to the south and spawning disturbances at the surface.

Maps like this one show the wind at the effective vertical half-way point of the atmosphere, where a barometer would measure 500 millibars of pressure. The lines on the map are height contours, which help to show how quickly the vertical thickness of the atmosphere decreases in the transition to colder, higher latitudes. Wind strength and direction are indicated by feathered arrows, where each feather is 10 knots, and a small triangular flag means 50. Note the symmetry of this graphic, and how clearly it shows weather events as linked features along a wave train. Cádiz is in the southeast corner of this drawing, just west of Gibraltar.

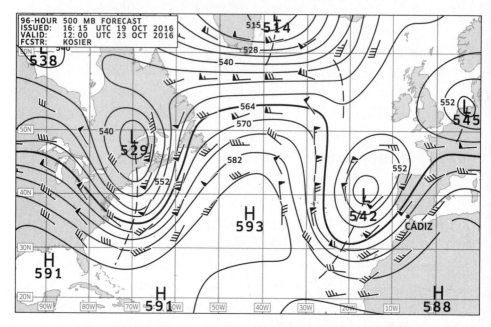

Fig. 9.2: A 500-millibar forecast map, showing upper air wind patterns
predicted for 12:00 UTC (Greenwich Mean Time) on October 23, 2016.
Our departure port of Cádiz is shown just west of Gibraltar.

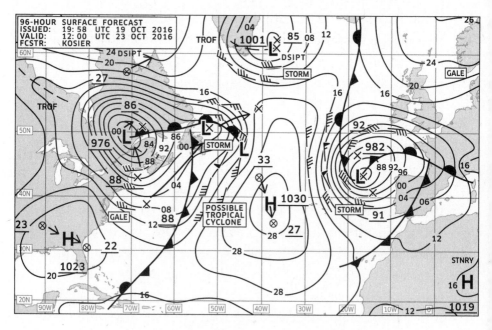

Fig. 9.3: A surface weather map of the North Atlantic,
also valid for 12:00 UTC on October 23.

The next map, Figure 9.3, shows conditions at sea level for the same interval. They are not peaceful. Note how the deep southerly meanders in the jet stream have created matching storms on both sides of the Atlantic: Cold air pushes south, creating instability as warm air is lifted and swept away by the strong winds aloft. A large mass of stable air in mid-ocean—the Bermuda-Azores High—is helping to steepen the pressure gradient surrounding each system, particularly in the eastern Atlantic. It was the rapid development of this storm and its predicted track to the south that drove our decision to return to Cádiz. On this map the changing values of barometric pressure are shown by isobars drawn at 4-millibar intervals. Note how the storms on both sides of the Atlantic are marked at the center by the letter *L* (for low) and a number indicating the central pressure of the system. A similar notation, *H*, is used for high-pressure systems. The boundaries between air masses are drawn as lines indicating fronts, with a sawtooth pattern for cold fronts and rounded lobes for warm fronts.

The strength of a jet stream depends on the temperature contrast between the air masses lying to the north and south of it. Jennifer Francis, a scientist at the Woodwell Climate Research Center in Woods Hole, has devoted much of her career to the study of this principle. More specifically, Dr. Francis is concerned about what will happen as the contrast diminishes. The whole world has gotten warmer, but the Arctic is warming fastest of all—as much as 5 degrees Celsius during the first twenty years of the new millennium. One culprit in this process is a positive feedback loop created by the melting of arctic sea ice: Warmer temperatures caused by an abundance of heat-trapping greenhouse gases create larger and larger ice-free areas, and the open sea surface—with its low reflectivity—absorbs up to nine times more solar radiation than ice. This has the effect of pushing temperatures even higher through a phenomenon called *arctic amplification*. On land, snow melts faster and sooner, leaving bare ground, which—like seawater—is a much more efficient absorber of solar energy than snowpack. The end result is a smaller temperature difference across the polar front, and a concurrent drop in the velocity of the jet stream—as much as 14 percent at certain times of the year.

According to Rossby's equations, a slower jet stream means slower-moving waves with greater amplitude. Troughs and ridges will push farther north and south, and stay parked for longer. This is one potential explanation for the temperature extremes that have occurred worldwide amid a general trend of global warming. The brutal North American winter of 2015 was accompanied by a deep and sluggish trough of arctic air parked for weeks over the continent. At the same time, a stubborn ridge delivered uncommonly mild temperatures to the West Coast. Europe endured lethally frigid temperatures in 2012, while the American winter set new precedents for warmth—all under the influence of sharp swerves in the jet stream.

These extra-long waves may help to block the east-west movement of surface weather, an effect that Jennifer Francis calls "stuck weather" in her discussions. Stuck weather could be blamed for the European super-summer of 2015, when athletes at Wimbledon were forced to play tennis in 96-degree heat and cities on the mainland set temperature records day after day. On the maps the cause was plain to see: a large northward bulge in the jet stream had allowed a huge, stable mass of North African air to park itself over Europe. There it sat and baked, while blocking the influx of cooler maritime air from the west. Weather analysts call this phenomenon an *omega block*, for the Greek letter whose shape it resembles on the page. Something similar to this happened early in the American summer of 2021, when an immense ridge of hot dry air took up residence over Washington State and British Columbia. A region accustomed to cool damp winds assumed aspects of the Mojave Desert. In Seattle the temperature rose to 108 degrees Fahrenheit. Desperate for relief, a friend I have there built an air conditioner in his kitchen: a wet towel hung from a broomstick, arranged in front of a roaring box fan.

"It's a square-rigged swamp cooler!" he exclaimed, his improvised relief ship sailing hopefully past his dishwasher, pulling heat from the room to evaporate water. The Moorish architects of southern Spain used the same principle to great effect in their castles, room after room kept cool by thin streams of water cascading through a series of fountains. In a disquieting glimpse at a potential future, the Seattleites learned to act like Andalusians—sitting quietly in their houses at noon with shades drawn, going forth only in the most pressing of emergencies. I have no idea if Kevin's family received respite through his inspiration, or merely entertainment.

THE year of 2021 also brought a happy announcement from NASA that the remarkable mission of their Juno spacecraft would be extended into 2025—promising several more years of miraculous images taken from its orbit over Jupiter. Jupiter's vivid horizontal bands are themselves jet streams, zonal torrents of gas produced by the same fundamental forces as on Earth: latitudinal changes in density set against the whirling framework of a turning planet. In this extraterrestrial case, racing clouds of ammonia and exotic hydrocarbons are suspended in an atmosphere of hydrogen and helium, far above a liquid surface. Among the jets are meanders and vortices, storms hundreds of years old and broad enough to engulf Earth in its entirety. Everything about Jupiter is gigantic. New data from Juno suggest that some of the Jovian jet streams may be as much as 1800 miles deep, with winds of 800 knots. Could one go sailing in this atmosphere, with the right vessel? Probably not. Nonetheless I wander off in thought to contemplate such things, at times when the sorting of earthly facts starts to feel too prosaic.

Mariners, I have found, share their confidences with remarkable candor. More than a few of my colleagues admit to episodes of a common anxiety dream, recurrent with remarkable consistency: They have run their ship aground. Worse, they have run aground so badly as to be sailing down a paved highway, swerving through traffic and looking for an exit back to water—knocking away signage and traffic signals with their mastheads as they pass underneath. I have had these dreams myself, set always on a certain road near my childhood home. With these visions I recall an even older series of dreams, reaching back well before I had any notion of my professional future as a navigator: I am in space, making an approach in some unknown craft to a planet that is unmistakably Jupiter, its brightly colored clouds swirling past my windows as I realize we have been brought to a halt by the force of the wind. Someone I don't recognize approaches me with a question about our plan. I can't answer them for sure. I am not frightened, but keenly aware that we cannot land until we have reached our destination.

10

CLAMBAKES OF ANTIQUITY

At the market in Tahiti whole fresh tuna are piled near a doorway, stacked in shiny pyramids like artillery shells by a gun battery. The smell of vanilla drifts around crowded tables where vendors sell produce and bottles of monoi—a clear scented coconut oil that turns instantly to a milky solid when I bring it home to Maine. For a few rare days in high summer it will revert to its lucid state, my wife calling me to bear witness.

"Quick!" she will exult. "We're back in Tahiti."

Upstairs from the fish stalls a Marquesan tattooist named Efraima is performing his singular work on a steady stream of our crew, some of whom have made appointments weeks in advance. Tahiti is our only port of call where staff demand time off from their workdays for tattoo appointments. The requests are granted. How can a morning of greasing winches or chipping paint compete with such attractions? Once inked, the crew return to work, the latest in the long line of mariners to bear dermal imprints of their adventures in the South Seas. Efraima's studio is deep in an alcove, past shops filled with flowered shirts and black pearl jewelry. There's no flash art on the walls, or catalog of flaming skulls and bleeding hearts for his clients to choose from. Each design starts instead with a conversation—which for those not fluent in French or Marquesan is limited to halting queries, gestures, or perhaps reference to a piece that someone else on the crew is already wearing.

"Like she has. An armband, with an, um, sea turtle in the design. Comment s'appelle *turtle*, en français?"

"Tortue."

"Merci, bon! Un tortue, s'il vous plait . . . And also—um—an anchor. Comment s'appelle *anchor*, en français?"

"Anchor."

Efraima abides. I wonder if by now we have put his children through college. After a quick pen sketch done right on the skin, he goes to work. Only when the ink starts to flow do the actual designs emerge, large motifs embedded with details taken from a stored vernacular of traditional themes. He inks bold animistic silhouettes, inlaid with fractal filigrees like the church carvings we saw at Nuku Hiva: A manta ray. A frigate bird, tail streaming, watching over a canoe under sail. At work Efraima is relaxed, his giant hands dexterous. There's music playing that sounds a little like reggae with French lyrics, and his head nods in time as he reaches back for the ink pot with the tip of his needle. The music is catchy, then endless, as the loop repeats three times while I'm there.

It is January of 2013, my third visit with the ship to Tahiti. Buildings around the harbor meet the water in a maze of cargo wharves and fish piers, backed by the sprawling center of town. The green bulk of the island looms behind, roads and buildings ascending the lush landscape until the ground becomes too steep for them to climb. Rain is coming down in sheets and pelting off of the rooftops like popcorn. The streets are coffee-colored rivers, splashed up into gritty curtains by passing cars. A phone rings and Efraima briefly sets down his tools to answer. Just outside, the outfall from some unseen gutter pours past our window, racing for ground like a startled snake.

Back on the ship the crew have abandoned all hope of doing any work. There are awnings improvised from tarpaulins, set up in a doomed attempt to keep the deck at least semidry. Overwhelmed, the tarps hold water like balloons and drop it all at once in in great cataracts, loud enough to wake you at night. This is something beyond rain, a pure wall of water that met us a day ago as we made our way cautiously into port. To the south were the misty green mountains of Mooréa, fringed with flickering surf, and then suddenly nothing, the visibility gone to zero amid a deluge dense enough to fill your coffee cup while you watched. We were approaching the harbor entrance—or so the chart told us—but the world in every direction looked the same. This time there was no sign at all of surfers, waterfront structures, or anything but a filmy gray curtain hung just beyond our bowsprit. It was calm, at least—perfectly windless as though the sky had for the moment converted all its energy to water. Someone in the chartroom made futile adjustments to the radars in an attempt to see anything at all, reaching for the handset as the radio came to life.

"*Robert C. Seamans*, you are clear to enter. Welcome back to Tahiti."

It was the harbor control officer, cordial as always but with noticeably reduced enthusiasm. The chief mate came back from the foredeck to talk, looking like he'd just walked under a waterfall.

"What do you want to do?" he asked.

I thought for a minute. It was raining harder than I'd ever seen it do in my life. How long could this go on? I wondered. How much water could there possibly be in one atmosphere?

"Let's go in another half mile," I said. "If we can't see the markers from that distance, we'll stand off and wait."

"OK."

He walked back up to his station on the foredeck with nothing in his movements to indicate that it was anything but a warm sunny day. We inched ahead. The rain eased from deluge to mere downpour, and the breakwater appeared slowly to port like a smudge on a wet windshield. We entered the harbor basin to find that everyone else had apparently been waiting for exactly the same chance to make their move. There was a tugboat crossing in front of us, a yacht shifting berths, and the morning ferry to Mooréa charging casually off the dock as ferries do. A Parisian traffic jam with very large cars, the radio lively with jabber as everyone pressed their own navigational agenda.

⌇

I T is still raining three days after my visit to Efraima, amid the arrival of news that a tropical cyclone is forecast to form near Samoa and head our way, sometime in the next forty-eight hours. "Where is Samoa?" someone asks me. We are waiting in a car outside the harbor office, and without a chart for reference this is not a straightforward question to answer.

"About 1500 miles west of here," I say.

Miles across an interstellar expanse of warm ocean, I think, one small world among many at the heart of Polynesia's sprawling galaxy. Samoa, the Cook Islands, Tonga, all within easy reach from here across the trade winds. Farther west still, past the ethnic boundary of Polynesia, there is Fiji—near the longitude of New Zealand and the first place large enough to spot on most maps without a magnifying glass. Draped across all this is the South Pacific Convergence Zone, a branching dogleg of the ITCZ that is large and nettlesome enough to merit a name of its own. It is OK to admit that you've never heard of the South Pacific Convergence Zone, but if you are an occupant of this great spread of islands and water known as Oceania, it will be a major player in your plans. Here humid trade winds crowd into the warm western Pacific, buffeted by seasonal movements of Australian air and the outflow from migrating high-pressure systems near New Zealand. During active periods the place is a morass of convection, a hatchery for storms as clusters of squalls coalesce into larger features—depressions, tropical gales, and in each austral summer several full-strength hurricanes.

Cyclones rotate clockwise in the southern hemisphere. A tropical storm coming from the west will thus have north winds on its near side, perfectly arranged in our case to wreak havoc on Tahiti's exposed main harbor. It is too soon to be truly worried—no, that is a lie. I am in fact, behind my commander's calm facade, deeply concerned, feeling the brief frisson of dread that I suspect all captains do upon seeing the dotted line of a storm track wandering their way. This particular storm is Cyclone Garry, expected to reach hurricane strength about halfway along its route from Samoa to Tahiti. Garry is for the moment moving slowly, now perhaps three days distant.

When I get back to the ship, one of the scientists comes asking for money to buy soursops at the market. A fruit like no other, they appear only fleetingly amid the abundance of taro and papayas, their rough green skins like alligator hides. I give standing orders to buy them on sight. On deck an awning is outdone by its load of rainwater and lets go with a thunderous splatter. I open the safe, imagining Papeete in a tropical cyclone—the storm surge overrunning the breakwaters and the harbor turned to a debris-filled ocean. I picture waves breaking in the park beside us, where families gather around the food trucks for crepes and poisson cru each evening. I want no part of it.

The currency of Tahiti is the Polynesian franc, beautiful money printed on multicolored bills emblazoned with sea turtles, flying fish, and sailing canoes. I count a crisp stack from the cash box and hand them over.

"Hurry back from the market," I tell the scientist. "We might be leaving soon."

~

THE South Pacific enjoys middling status among the world's hurricane basins—well behind the northwest Pacific, where the local typhoons make up a third of the tropical storms born annually on the planet. South of the equator such storms are known simply as *cyclones*, and while they tend to be smaller and less frequent, their behaviors are alarmingly erratic. I grew up watching hurricanes in the North Atlantic, their movements narrated by the calm voice of the National Hurricane Center. The NHC is a veritable control tower of storms, loquacious, if not omniscient. Here in the vast South Pacific, cyclone forecasting is shared between the capable but less media-rich weather services of Fiji and New Zealand—and the storm tracks, to anyone familiar with the smooth sinusoid trails of Atlantic hurricanes, evoke the footprints of grazing sheep.

South Pacific cyclones form amid the instability of the South Pacific Convergence Zone, often following its axis to the southeast and sometimes doubling back before escaping into the prevailing westerly flow of the higher latitudes. In February of 2016 Cyclone Winston formed near the Solomon Islands. It drifted south and then east to take two swipes at Tonga, next reversing into Fiji as the most powerful storm ever to make landfall in the southern hemisphere. After taking forty-five lives and doing 1.5 billion dollars' worth of damage on the island of Viti Levu, Winston blundered south again, then west to finally expire near Australia.

Cyclone Gita formed near Vanuatu in February of 2018 and drifted a thousand improbable miles east, grazing Samoa before making a wide U-turn to strike Tonga. I had watched Gita with some anxiety as it continued west, nearly to Australia before looping back toward New Zealand, where my ship was hidden far up an inlet and secured stoutly with a spiderweb of docklines. In the end Gita delivered wind and heavy rain to the coastline south of our location but spared us from grief, save an afternoon of dark squalls and a strong dose of adrenaline.

J ANUARY is considered the rainy time of year for French Polynesia, though this one is particularly rainy by any measure. January is also tropical cyclone season, but it's rare for storms to come as far east as Tahiti, whose longitude is roughly comparable to that of Hawai'i. For us, however, things are beginning to show that it might not be an average year. On the climate pages, an obscure benchmark number called the Oceanic Niño Index is tipping into positive territory, and in response forecasters are discussing the possible advent of an El Niño event in the near future.

El Niño is the peak of a multiyear oscillation across the tropical Pacific, driven by a joined interaction between ocean and atmosphere. Imagine water sloshing back and forth in a bathtub. In this case the bathtub is Earth's largest ocean, with South America at one end and New Guinea at the other. That's eight thousand nautical miles—a third of the planet's circumference—with less dry land than the state of New Hampshire. Under normal conditions warm currents in the Pacific are driven west by the trade winds, crossing the length of Polynesia to collide eventually with the dense archipelago spread between Australia and Thailand. Delayed here in its travels, the water slows down and starts to bake in the sun. This is the Western Pacific Warm Pool—the warmest bit of ocean on Earth, with surface temperatures as high as 30 degrees Celsius. It's a big hill of warm water, piling up faster than it can spread back out again. Satellite measurements show that the surface of the warm pool can be as much as seventy centimeters above mean sea level in the eastern Pacific. In an average year, evaporation over this superheated mega-puddle lifts enormous amounts of water vapor into the atmosphere, making rains that irrigate the verdant forests of Borneo and sustain the food crops of Southeast Asia. Thus dehumidified, the air is then carried at altitude back to the east, growing steadily cooler and drier before settling eventually again to earth.

All this makes for a big difference in climate between the two sides of the Pacific basin. The southwest Pacific is wet and cloudy for much of the year, while the eastern parts tend to be dry and fair. Much of South America's west coast has an arid landscape, similar to Baja California or southwestern Africa. Here the trade winds push seawater away from the shore, where it is replaced by a cold, nutrient-rich flow from underneath through the process of upwelling. This is the origin of the cold tongue that we found on our way to Nuku Hiva in 2007, its plume of richly productive water spreading offshore to nurture the bountiful fisheries of Peru and Chile.

The Peruvian fishermen were never sure why, but every few years the trade winds would slacken, the coastal waters would warm, and their fish would vanish as the upwelling cycle collapsed. Seabirds died in droves. Ashore, humid west winds made rain in the desert, sometimes to the point that flooding became a problem. This phenomenon tended to peak late in the year, around Christmas, and because of this they named it *El Niño*—the Christ Child.

As it turns out, El Niño is produced by a periodic slumping of the Western Pacific Warm Pool, often preceded by a year or two of above-average trade winds. In simple terms, there is a limit to how big a pile of warm water you can make in one place, and eventually the whole thing wants to slide back the way it came. This cycle is regulated by a dense combination of oceanic and atmospheric factors—the math gets complicated quickly—but the result is a slick of warm water spreading back into the central and eastern Pacific. This means more rain in certain places, and for Peru a disturbance in the coastal ecosystems that are critical to fisheries. The vast relocation of energy and moisture has a worldwide ripple effect on atmospheric circulation, which is why the rest of us have learned to care when it happens. An El Niño year may bring drought to Indonesia, hurricanes to Tahiti, floods to California, and a retinue of other impacts that get progressively harder to pinpoint as you move farther downstream from the cause.

There is an opposite to El Niño, called *La Niña*. In La Niña years, extra-strong trade winds drive the warm pool farther west than usual, and increase upwelling off of South America. The tongue of upwelled water becomes expansive, reaching out along the equator and yielding conditions that can feel anything but equatorial. We didn't know it at the time, but our chilly crossing aboard *Robert C. Seamans* in 2007 turned out to be the product of a La Niña year—an ample, if unsatisfying, explanation for all the hats and mittens broken out even as the latitude counted down to zero.

The atmospheric arm of El Niño involves the *Walker circulation*, a set of convective cells over the Pacific that are driven by differences in sea and air temperatures along the equator. The juxtaposition of cooler air near South America and warm air over the Western Pacific Warm Pool creates an east-to-west pressure gradient between Peru and Oceania. Cool dense air drains down off of the Andes and is carried west by the trade winds, accumulating heat and water vapor along the way. Over the warm pool, it is buoyed aloft by convection, releasing moisture as it rises. High in the troposphere, the lifted air then slides back to the east, cooling and sinking eventually back toward the surface. A standard weather schematic of the tropical Pacific would show dry and settled conditions to the east, and a constant mass of rain clouds to the west, where the air is pumping steadily aloft. Surface pressure over the warm pool stays perennially low, a positive feedback mechanism that encourages the trades to blow even harder.

In an El Niño period, the overfilled warm pool sags eastward and the cells of the Walker circulation move to follow the migration of warm water. The middle Pacific sees more rain, and the overall circulation gets less powerful, since the east-west differences in temperature and pressure are less profound. Tahiti gets wet, Peru gets fewer anchovies, and without the rainy western pillar of a Walker cell parked overhead, Indonesia dries out like a prune. Another earmark of El Niño are wind anomalies, where the trade winds fade and may actually reverse themselves— blowing from the west instead of the east as their normal mechanism breaks down.

Such westerly winds doubtless helped sailing canoes reach the distant corners of Polynesia in the heyday of Pacific voyaging, centuries before the maritime expansion of Europe began. Consider the atoll of Tongareva, in what are now the Northern Cook Islands—six hundred miles below the equator and seemingly in the middle of nowhere, except that Tongareva happens also to sit on a perfect horizontal line between the Solomon Islands to the west and the Marquesas to the east. All that stands in the way are several thousand miles of empty ocean, easily traversed if you have the wind in your favor and can keep track of your latitude. In his book *Pathway of the Birds*, Andrew Crowe tells a story of Portuguese sailors who visited this place in 1606 and found a

Neutral Conditions

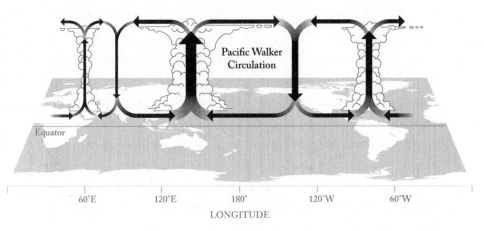

Fig. 10.1: Pacific Walker circulation in neutral climate conditions.

El Niño Conditions

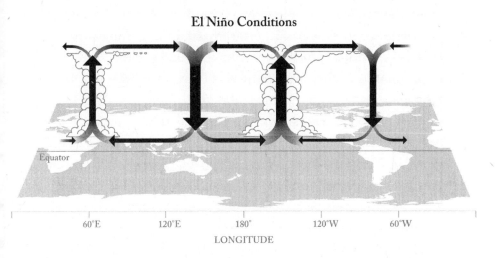

Fig. 10.2: Pacific Walker circulation in an El Niño year, with the main area of convection shifted east toward French Polynesia. In these conditions, sporadic westerlies may replace the normal trade winds, an opportunity for sailing east in the tropics as sinking air over Oceania leads to a collapse of normal wind patterns.

thriving population, living in well-built homes with their heavy canoes beached nearby—no doubt awaiting a timely circumstance for departure.

Such stories play in my own imaginings of how this must have looked in real time. Aware of their place in a world of water and islands, the Polynesians surely knew of transient climate events, their cycles frequent enough to be experienced firsthand by each successive generation of navigators. I am picturing two men on a beach, five hundred years ago. Perhaps they are watching the east wind start to die or noticing new birds browsing the scrub and skimming over the reef. The sky has altered subtly in its coloring, reflected back in the riffled surface of the lagoon.

"It ought to be time soon," says one. "Got your boat loaded up?"

"Just about," says the other. "My brother-in-law has my garden tools, but he has promised to bring them back by tomorrow."

The first people of the Pacific began in the west, trickling first down the pendulous topography of Southeast Asia and reaching eventually as far as America, or nearly so. In their minds the islanders held maps of their own making, bound together by different notions of what was connected and what was not. The back-and-forth spread of durable things that they left along their way—tools, place names, crop species, and DNA—proves a repeated pattern of travel along these routes. A Tahitian plant grows in New Zealand. A stone adze from Hawai'i is found in Nuku Hiva. Garden implements taken away by family members on canoe trips were perhaps in fact not lost forever.

I first considered Tongareva as a destination late one night amid what may still be the scariest moment of my career. A shipmate critically ill, the ship in mid-passage from Hawai'i to Bora Bora. On the satellite phone, a crackly voice, weak but eager to assist:

"Yes, *Robert C. Seamans*—what is your nearest airstrip?"

Near is a relative term, at such moments. There on the chart margin, a mere 450 miles away, was Tongareva—a place I'd never been or truthfully even thought of until then. Who would we find there? I wondered. In the end we did not go, but the vision of this briefly noticed location remains embedded in my recollections. Imagine a long night drive across the empty middle of America, perhaps Kansas. Pure darkness but for a single passing cluster of lights on a rise far away. A gauge flickers on the dash and you look toward the lights, wondering briefly who lives there and if they might help.

Sɪʀ Gilbert Walker didn't know anything about the Pacific Ocean when he took a job at the India Meteorological Department in 1904. Walker was a mathematician, not a meteorologist, but his predecessors had the foresight to realize that math would be needed if they were going to understand why the Asian monsoon had disastrously failed to develop during several years in the late nineteenth century. These are the same monsoons that brought Arabian dhows back and forth from Zanzibar to Calcutta, a seasonal cycle of wind between the continental and maritime environments of South Asia. In spring and summer, the air over land warms and rises. This allows moist oceanic air to slide in underneath, bringing rain to regions that may otherwise be arid for much of the year. Arizona has a monsoon also, but it is small beer compared to the one in Asia, where an entire continent relies on seasonal rains to support its agriculture. When this failed to happen in 1877, the result was catastrophic famine.

Walker reviewed decades of barometric data from all over the world and noted that there were distinct seesaw patterns in surface pressures across Earth's three major ocean basins—the North Atlantic, the North Pacific, and most significantly the South Pacific, where he saw that periods of abnormally high pressure over places like Australia coincided with lower pressures to the east, and vice versa. Further, he noted that the extremes of what he eventually named the Southern Oscillation correlated with rainfall and temperature trends in a diverse range of locations—India, Chile, North America, even Africa. Walker did not solve the whole puzzle, but the statistical tables he built for predicting rainfall have stood the test of time and were the first quantitative attempt at climate-based, long-range weather forecasting. This work developed eventually into the Southern Oscillation Index, a forecasting tool now standardized to compare surface pressures between locations in Tahiti and Australia—opposing seats of the El Niño climate seesaw. Sir Gilbert did not fully answer the monsoon question, but it turns out to have a direct link to the Southern Oscillation. The elevated pressures in the western Pacific during an El Niño episode lead to drier air and weaker winds over the Indian Ocean, which in turn means less moisture migrating ashore to make rain.

Jacob Bjerknes gained early renown as author of the 1922 Norwegian Cyclone Model—a seminal finding that anchored the study of temperate weather systems for most of the next century. That achievement should have been enough for any career, but four decades later he was also first to recognize that Walker's Southern Oscillation was coupled to El Niño through a system of direct interactions. In their love of acronyms scientists have coined the singular title of ENSO (El Niño–Southern Oscillation) to describe these linked phenomena. Bjerknes gave the Walker circulation its name and saw how it could act as a forcing mechanism for South Pacific climate systems. He used historic data to lay out a set of baseline conditions for the Pacific, and then used measurements from the El Niño period of 1957–58 to show how high surface pressures in the western Pacific had coincided with a failure of the trade winds and a shutdown of upwelling off Peru. With no easterly winds to restrain it, a great lake of warm water had made its way back toward America, leading to an abysmal year for the anchovy fishermen. In contrast, Bjerknes also noted how stronger-than-average trade winds could yield La Niña episodes through what is essentially a magnification of the normal—boosting the upwelling of cold water off South America and increasing the accumulation of energy in the warm pool far to the west.

The winter of 1957–58 saw stronger storms in the North Pacific, along with a stubbornly entrenched low-pressure regime over western Russia. Proposing a downstream connection between El Niño and global weather patterns, Bjerknes blamed all of this on the overheated eastern Pacific: All that extra energy in the ocean, he posited, was heating the atmosphere at a brisk rate, and the subsequent surge in convection was pumping warm moist air up into the jet stream. This made for stronger cyclones to the north and aimed a fire hose of tropical moisture at the US West Coast and the Gulf of Mexico.

For all that he discovered, Bjerknes lacked the data to determine just where the tipping points were between the two halves of the ENSO cycle. How, exactly, did the system reset to neutral after an El Niño had run its course? And when thereafter would it begin tilting toward its other extreme? Sorting these details took up most of the next two decades, as scientists improved their understanding of how energy is transferred between ocean and atmosphere. To a point, it was a bit of a chicken-egg problem, since any oscillating system is a continuing cycle of accumulation and rebound. In 1975, a researcher named Klaus Wyrtki noted that El Niño episodes were often preceded by a period of strong trade winds and an "overcharging" of the Western Pacific Warm Pool—which in turn produced an easterly countercurrent of warm water in the form of something called a *Kelvin wave*. This mass of warm water bouncing back toward South America would weaken the trade winds, making it easier still for the warm wave to roll farther east. Wyrtki used data from tidal gauges to support this theory of a tilting sheet of water coursing back and forth across the ocean, noting that sea level heights off Peru could rise by as much as twenty centimeters in an El Niño year. Eventually, enough warm water would accumulate to reignite the trade winds, and the system would begin to slosh back toward equilibrium. A full run of this process took anywhere from six to eighteen months.

In 1986 a team at Columbia University achieved the first mathematically modeled forecast of an ENSO event, a feat that they managed to duplicate with another successful prediction in 1991. A lot of work since then has gone into refining the models, though it is still an inexact science. In 2014 scientists at NASA and NOAA—partner organizations in climate forecasting—felt that Pacific surface temperatures were high enough to predict an incipient El Niño with near certainty. Despite this, the atmospheric mechanisms failed to respond, and the result was unimpressive. In the following year of 2015, NASA declared a looming El Niño "too big to fail," and this time they were right—the subsequent event met or exceeded most previous records.

In spite of such early warnings—perhaps due to cries of "wolf" in the preceding year—the Indonesian weather agency had difficulty getting its message across to farmers about the need to exercise caution in their fields. The dry season is the traditional time to burn back brush at palm plantations, and in 2015 the process worked a little too well, as controlled burns turned to conflagrations amid the drought.

The perturbations of El Niño propagate downstream as a series of ripples in the overall flow of the atmosphere. Bjerknes's sample year of 1957–58 showed a typical pattern of secondary effects—heavy rain and snow in the American West, warm conditions in the western Atlantic, and a harsh European winter as a

Neutral Conditions

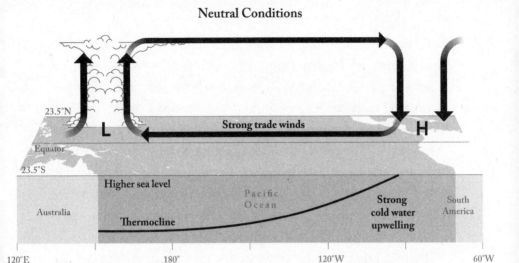

Fig. 10.3: During neutral and La Niña years, strong trade winds produce a higher sea level in the western Pacific. Here the air temperature is warm, and the atmospheric pressures are low at the surface. The thermocline (the transition layer between warm surface water and colder deep water) is steeply sloped from east to west, with colder water rising to the surface off South America.

El Niño Conditions

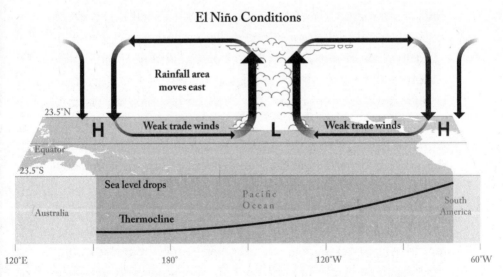

Fig. 10.4: In an El Niño year, the sea level drops in the western Pacific, as the trade winds weaken and warm water "sloshes" back across the Pacific toward South America. Atmospheric pressure over Oceania increases as the main area of convection follows the warm water east over the central Pacific.

low-pressure trough parked itself over Russia for weeks on end. That pattern manifested again in the winter of 2015–16, when the eastern US experienced rare December tornadoes, the West got hammered with rain and snow, and Europe was battered by a series of intense maritime storms that brought record rainfalls to parts of the UK and Ireland. As proof that few things are bad news for everyone, El Niño years are generally connected with fewer North Atlantic hurricanes, as the strengthened subtropical jet streams increase wind shear and tend to destroy the vertical symmetry that tropical storms require in order to grow.

There is great speculation as to how all this will be affected by a warming climate, but historically ENSO is nothing new. Scientists at the University of Maine have used the refuse from millennia of Peruvian clambakes to draw a timeline of El Niño into the past. It turns out that mollusks are an excellent proxy for climate. Many can't tolerate big swings in sea temperature, and the average bivalve lacks the mobility to relocate when the weather becomes unfavorable. The Maine team looked at shell heaps left by ancient coastal societies and noted that certain species had vanished in favor of more heat-tolerant breeds during specific periods. Inland, they used the geological records of flood debris to corroborate these findings, while other studies used the preserved strata of highland plant materials as a similar proxy. Flood deposits marked El Niño years, while layers of upland peat indicated the prolonged periods of cool and mossy conditions coincident with neutral or La Niña intervals.

All of this led the Mainers to conclude that our modern version of El Niño has been in action for about three thousand years. Before that, episodes were apparently rarer—once in a human lifetime, perhaps. And prior to about six thousand years ago—the rough halfway point of the present-day Holocene epoch—it appears that the tropics were warmer overall, and with fewer fluctuations. The archaeological records in Peru show a pattern that seems to mirror this variability. The temple mound settlements of western Peru are the oldest known cities in the New World, and they flourished in the period that preceded the onset of modern El Niño conditions about three thousand years ago. After that point, the one-two punch of periodic deluges and bad fishing might have conspired to drive them elsewhere. The last temple mound city to survive in the region under study was at a place called Manchay Bajo, where archaeologists found evidence that levee walls were being built during its final years in an attempt to counter the effects of flash flooding.

~

O N the south side of Tahiti, we bring the ship to a deep lagoon with good anchoring bottom and plenty of room to maneuver if things get lively. Here the mass of the island's volcanic dome gives fine shelter to winds from the northwest. This is where the French warships came long ago, before construction of the commercial harbor at Papeete. You can still find their massive concrete moorings along the shore, splashed with graffiti and strong enough to hold an aircraft carrier. I figure that if need be, we can take our lines to one of those and sit securely through quite a lot of wind. Tapueraha was Tahiti's first deepwater port, but there is nothing here now, save for a line of tattered palms and a rickety roadside bar that is closed every time we check. From our berth in the anchorage we see cars roll past, flashing in and out of sight between the trees. By the ruins of an old marina a local family is camped under a tarp, their children bounding off a crumbling pier and into the water. The rain continues, modulating slowly between a light drizzle and a biblical downpour. We scarcely notice after a while.

We do projects, play cribbage, and venture out to visit the

famous surf spot at Teahupoo, not far up the narrow road. For a month each year this place becomes a madhouse, as pro surfers come to cheat death on the giant winter swell, but for now it's just a fishing village. Dogs circle cautiously, and someone waves from under the palm-thatch roof of a carport. Past here the road ends and there is nothing but the jungled shore of Tahiti's untracked eastern appendage. We wait thirty-six hours at Tapueraha, long enough to be sure of where the storm is going. In the end, Garry never finds us. It makes its way east and eventually south, deflected into the Southern Ocean, as Polynesian cyclones sometimes are.

Feeling better about the future, we set sail again on what proves to be a long soggy loop through the western Society Islands— Mooréa, Huahine, Bora Bora with its photogenic spires of stone, and Raiatea, thought by some to be the point of departure for the final great round of Polynesian sailings a thousand years ago. Through it all the rain chases us in waves, returning like clockwork after each deceptive interval of sunshine. The wind is fitful, gusty with the passage of squalls, and then gone altogether, the sails hanging wet and slack. There is a steady swell rolling in from storms far away, its long relief visible on the smooth water as floating seabirds rise and disappear in the alternating peaks and valleys. The soursops from the market grow ripe and are devoured, spiny green globes with hard black seeds and a filling like vanilla custard. Recovery to normal fruit-eating is not possible once you have had such things.

With our oceanographers I review the data and agree that we are without doubt observing the classic harbingers of an El Niño event—elevated water temperatures, stagnant trade winds, wet conditions, and the threat of cyclones well east of their normal range. A month later reports from NOAA and NASA begin to waffle. Maybe not this year after all, they say. The trade winds reappear—fitfully at first and then a warm steady breath of re-stored momentum. The various climate indicators dip back into neutral territory, and warnings are called off. The scientists take all this in stride, it being part of their profession to understand that there is in fact no such thing as a classic harbinger. It is sim-ply necessary to realign one's conclusions with the latest data at hand. There is, I learn, already a name for the sort of ENSO head-fake we have just been subjected to. Scientists call it "La Nada."

11

THE ICE DRILLS

M ariners step back from the daily exigencies of weather and into the world of climate when they lay out plans for future voyages. How easy or difficult might one expect a passage to be on a given ocean at a certain time of year? Is it better to run offshore or hug the coast? Will the trade winds, timeless engine of sailing ships, be reliable or elusive? And how likely is it that a hurricane will come along and turn a carefully laid plan on end? Much of this statistical information is embedded in a fascinating publication called the *pilot chart*—a bundled sheaf of pages showing, month by month, the historical trends of wind, temperature, ice, and storms for each ocean region. You don't navigate on these charts, but their distinct iconography will tell you what your ship might expect to encounter once the harbor buoy is left astern. This is the essence of applied climatology—the use of long data averages to guide decisions, just as maps from the old Weather Bureau once advised farmers in Iowa on when to plant their corn.

The father of the pilot chart was an American named Matthew

Fontaine Maury, a lieutenant appointed in 1842 to run the US Navy's nascent Depot of Charts and Instruments. Injured in a stagecoach accident and unable to sail, Maury threw his formidable energy into building a record of winds and currents for the world's oceans—a sailor's atlas for passage planning, built with years of data taken from ships' logs. Like Beaufort and FitzRoy in England, Maury was a leader in standardizing the format by which ships at sea recorded the weather, offering in return free copies of his charts to anyone willing to participate in his observational system. By 1861 he had developed a global suite of his signature products, including special editions for monsoons, ocean currents, storms, rain, and whaling.

The ship captains of California's gold rush used Maury's work to plan their routes and were rewarded with greatly improved results: By 1855 the average time for a sailing voyage from New York to San Francisco had shrunk by nearly 30 percent, from 187 to 136 days. A record run of 89 days, set by the clipper *Flying Cloud* in 1854, would stand until 1989. This watershed of advancement was stalled at the start of the American Civil War by Maury's defection to the Confederacy, but his pilot charts lived on. They are still available in updated form from the National Geospatial-Intelligence Agency, a bureaucratic descendant of Maury's old office. The modern charts preserve Maury's concept of dividing oceans up into a grid, each five-degree square of latitude and longitude embossed with something called a *wind rose*—a representative dial of wind averages for the month in question. Other markings show the percentage of storms and calms in each square, the general path and frequency of cyclones, ocean currents, and—in polar regions—the annual extent of pack ice and icebergs.

Dr. Kevin Wood is a researcher at the NOAA Pacific Marine Environmental Laboratory in Seattle. A digital heir to Matthew Fontaine Maury, he spends days absorbed in preparing old ships' logs for the process known as *reanalysis*—a supercomputer-powered regimen whereby modern images of past weather events are made by feeding antique data into the latest forecast models. Imagine a set of digital graphics for the 1900 Galveston hurricane or the Thames flood of 1928, both now free for viewing on NOAA's immense website. In fact, with the right queries one can ask the server for a snapshot of Earth's atmosphere at any point reaching back to 1836. It is a climate time machine, with raw computing power used to make up for the relative paucity of data from years like, say, 1861—when a storm called the Expedition Hurricane nearly doomed a Union Navy assault on Confederate fortifications in North Carolina.

Scientists call this method *sparse-input reanalysis,* the layman's equivalent of which might reasonably be called *making something from nothing.* Full-color pictures of century-old events are rendered to a degree that no archival record can touch. The main input used in reanalysis is surface pressure—hourly barometer recordings, taken by uncounted legions of ship's officers reading the glass on their respective journeys across time and space. A humble and routine measurement on its own, barometric pressure seems able to stand as proxy for a host of other values if the appropriate gigaflops are applied.

I sailed with Kevin years ago on a tall ship voyage to the Caribbean after a long summer spent in the loud squalor of a Maine shipyard. Come October we finally swept the last of the sandblasting grit from our decks and escaped, still bending on sail as the day faded and the coast disappeared astern in long skeins of fog. For the next month we stood our watches on what I now understand was a routine fall passage to the islands: The long initial nights of seasickness and equipment breakdowns. A smart pummeling by a gale, the ship adrift in driving rain and then careening off to the south as the cold front blew past. One hurricane scare, fitful calms east of Bermuda, and at last the trade winds, their warm exhalation driving us along at silent speed for days on end, suffering forgotten. Through it all I found on my pillow each night a perfect spoonful of sandblasting grit—not real sand at all but a distinctly obnoxious product called Black Beauty, made with silica slag from blast furnaces. I blamed my cabinmate Kevin for this serial prank—unjustly, it would turn out, as weeks later I would find that a leftover pile of the stuff had collected in an overhead ventilator and was somehow sifting down bit by bit onto my bedding.

A remarkable thing about old shipmates is how they always seem to take your calls, no matter how long it's been. When I catch up with Kevin, he is reviewing ships' logs from a storm called Typhoon Cobra—a tropical cyclone that inflicted grievous losses on a US Navy task force near the Philippines in December of 1944. Cobra had been spotted in its developing stages by weather aircraft, but in a push to keep his ships on station at all costs Admiral William Halsey ordered them directly into the path of what today would qualify as a category 5 storm. The costs were substantial. On the decks of aircraft carriers, planes broke free and started catastrophic fires. Smaller ships, low on fuel, grew dangerously unstable, and some capsized. In the end 790 sailors and three vessels were lost, with dozens of others damaged so badly that they had to withdraw for repairs. Written in the terse prose of military records, the log entries are still shocking to read. Force 19, some say—several steps above even the expanded sixteen-point version of Beaufort's wind scale that the Navy has adopted for itself. Probably, Kevin speculates, this is what gets written when someone is on a ship's bridge, rolling back and forth through a 140-degree arc and watching their vessel get blown apart bit by bit—deck lockers, ammunition boxes, lifeboats, and then the boiler stacks, their half-tethered steel barrels careening around like the wheels of a runaway mangle.

Kevin supervises the American section of a campaign to convert as much old meteorological data as possible into digital form for reanalysis. Known as Old Weather, this project began in England and took root in the US after Kevin met its director during a conference in 2010. Since then he has been the admiral of a volunteer navy, its sailors busy at desks and kitchen tables transcribing numbers from book pages into spreadsheet cells. Herein, I realize, is the foundation of all science—the endless and meticulous sorting of information. For anyone with a love of detail and time on their hands, a bottomless vein is there for the mining: There are still voluminous military records from World War II, set in crisp type and perfect for beginners. Seasoned recorders may turn to the more challenging logs of whaling ships—their tall pages covered in lines of looping quill script, done in tiny cabins just feet away from the smelly chaos of roaring fires and bubbling try-pots.

⌒

I N graduate school Kevin looked at logs brought back by the ragged armada of vessels that flung itself into the trackless abyss of the western Arctic, during the frenzied years of nineteenth-century exploration. While it was more or less proven by the reign of Queen Elizabeth I that no practical northern sea route existed between the Atlantic and the Pacific, the hungry machine of British colonialism never gave up on trying to find one. The Royal Navy, desperate for a mission after Napoleon's defeat in 1815, threw ship after ship into the unmapped maze between Canada and the north pole, all in search of the chimerical Northwest Passage. None succeeded.

In the most famous of these doomed adventures, a commander named John Franklin departed from Greenland in the summer of 1845 with two ships, 129 men, and the full wind of England's mania in his sails. His vessels, the *Erebus* and the *Terror*, were purpose-built veterans of polar exploration, with massive iron reinforcement, retractable rudders, and propellers powered by steam engines—the nineteenth-century equivalent of warp drive. There were heated cabins, a library stuffed with books, and three years' worth of food preserved in newfangled metal cans. "This is an expedition too big to fail," someone surely must have said at the time.

A formidable task awaited Franklin's enterprise. The fractured upper regions of North America are nominally Canadian, but in fact remain one of the most sparsely settled and least-traveled places on Earth. Here a complex mix of wind and current drives ice endlessly back and forth through a sieve of islands sown like boulders between the mainland and the Arctic Ocean, their treeless shorelines unlit by sun for large parts of the year. Frozen nearly into stone by annual cycles of softening and refreezing, the ice is aggregated by wind into something called *pack*, a ridged hardscape that pushes south in winter and retreats by halting increments during the brief summers. About as yielding as concrete, the pack can trap and crush unreinforced ships with ease. Icebergs are another regional menace, giant blocks broken off from glaciers that roam like aircraft carriers amid the sea ice. Driven by ocean currents, icebergs roll along with enough momentum to scour grooves in the seabed when they run aground. As history has shown, you don't want to hit one. The pack slows the bergs down in winter, leaving them to escape in summer when the sea ice retreats and areas of open water form brief windows for navigation.

With great fanfare, Franklin and his ships sailed off into this labyrinth and disappeared utterly from view, leaving a mystery that would galvanize exploration for much of the remaining century. The first real clues came in the spring of 1854, when a Scottish explorer named John Rae met a party of Inuit hunters deep in what today is the indigenous province of Nunavut. The hunters told Rae of a winter four years previously when they'd come upon a group of about forty starving Englishmen blundering slowly across the snow. The leader of this desperate crew, a large man with a rifle, explained that they were refugees from a thwarted naval expedition. Their ships had been stranded by the ice, and they were headed south in the hope of finding game to hunt. The natives, understanding the realities of the food supply in that bitter landscape, no doubt nodded politely and gave the qallunaat a wide berth. When the same hunters passed that way the following spring, they saw that the doomed whites had not gotten much farther. Amid the bodies were signs that some had taken to the desperate last resort of cannibalism in their final days. And along with the bones were the items that the lost men had felt essential to carry on their way—ebony pencil cases, gold pocket watches, and embossed English silverware.

The search for Franklin became a Victorian version of the space program. By the time the fate of Sir John was fully unraveled, a series of grueling explorations had filled in the empty map of the Arctic, proving that it was at least theoretically possible, if not well-advised, to cross north of Canada by water. This said, it was not until 1906 that the Norwegian Roald Amundsen actually did so, in a small sloop called the *Gjøa*. It wasn't easy. Roald and his crew picked their way through a chain of shallow side passages no big ship could have managed, close along the mainland shore. They spent two winters frozen fast before finally emerging somewhere north of Alaska. Amundsen himself then set out on a five-hundred-mile ski trip to transmit news of his success by

telegraph, claiming a prize that was as much about winter camping as navigation. It would be another forty years before a Canadian named Henry Larsen (another Norwegian, in truth) finally
got through in a single season, making the passage in a schooner
small enough to cross shoal water and dodge inshore to escape the
pack.

In his research, Kevin Wood was interested to know whether
the travails of these driven souls could—as many thought—be
connected climatically to Europe's "Little Ice Age," a period of
particularly cold winters that went on intermittently for several
centuries beginning in around AD 1400. One can blame the Little Ice Age for the failure of Norse colonies in Greenland and
Napoleon's defeat in Russia. Kevin reviewed the records of forty-
four expeditions altogether, volumes filled with names guaranteed to thrill anyone who's ever looked at a map of the Arctic. In
flipping through these hallowed pages Kevin arrived at some re-
visionary conclusions: Measured against the larger data record,
the Arctic of Franklin's era in fact appeared quite similar to that
of later times, at least up to about the year 2000. While nineteenth-
century Europe might have been bitterly cold, the ice conditions
in the Canadian Arctic were just average. And while the Northwest Passage has itself become something of a universal metaphor
for doomed quests, Kevin noted that many of the British explorers had come tantalizingly close to success amid their failings—
some within one hundred miles of triumph. A nineteenth-century
navigator gifted with GPS and a diesel engine might well have
enjoyed a better outcome.

❧

I N my own youthful voyages to Greenland, I allowed myself to
feel some small thrill of kinship with the earlier explorers of this
enchanted landscape. It was hard to think while sailing amid all

that rock and ice that the place had ever looked any different. We left Maine in late June, a point at which the pilot charts showed that the summer retreat of sea ice would likely be well underway. After a short stop in Newfoundland we sailed due north, up the 53rd meridian on a route designed to carry us well away from the Labrador coast—where a cold current generally keeps the ice around longer. Historically the probability of gales along this route was listed as low in July—between 2 and 6 percent—though we would soon receive proof that *rare* does not mean never. I pondered this lesson in statistics as the wind howled and the rain pelted down, regarding the lonely ocean and thinking that perhaps some data had been missed. How many gales blow here, I wondered, with no one to count them?

From our landfall after the storm we made our way up a black-and-white coast of snow-folded mountains, the water dotted with ice and hidden periodically by long slivers of mist. Night was replaced by the half-light of arctic summer, a charged gloaming that defied sleep and made time itself hard to judge. People rose at all hours, uncoupled from the clock and energized by the promise of arrival after a long, tedious passage. There were whales everywhere, along with watchful seals and fulmars—stubby seabirds with economical wingbeats and thick plumage built to keep the heat in. Approaching the Greenlandic capital of Nuuk, a surveyed channel ran inshore through the Kookøerne—a broad rockbound archipelago scattered out into the surrounding sea like spilled peas. To either side of us between the islands was just a blank white space on the chart, as though to suggest a place that nobody had ever visited. Some work remained for explorers to do, this seemed to indicate.

"These are Danish charts," the helpful harbormaster told me. "The best in the world. Everything is just where they say it is. And if you see nothing . . . well, you are on your own.

"Oh," he said, in an afterthought, "we have no navigational buoys either . . . because, well, the ice . . ."

We were in Nuuk for about seventy-two hours—a typical town of the modern Arctic, with drab concrete apartment blocks and a tidy harbor, shipping containers stacked in boxy rows by warehouses on the wharf. In the endless summer light young mothers pushed strollers past our berth at four a.m. and taxis circled the town square as though in Copenhagen. There is nowhere that you can actually drive to from Nuuk, or any of the other dozen or so main settlements in Greenland, but that has done nothing to stem the rising tide of vehicles. I was reminded of fishing villages that I knew from home, the tiny wharves crowded with new trucks and backyards filled with their rusted forebears. The Greenlanders still travel mainly in small boats between villages, or by dogsled in winter—a much-awaited time of year in which the bays freeze into highways and make such things possible. In the northern towns we would find the people outnumbered by dogs—rangy feral beasts, not for petting, tied outside every house with their idle sledging gear stored nearby.

The harbormaster made space for us along the wharf among the throng of fishing boats, each in port only briefly to unload their catch and replenish fuel before returning to work. My brother, our cook, was elbow-deep in a pile of shrimp we'd been given by the vessel alongside us in a characteristic gesture of welcome. Their captain led me up into his wheelhouse, a small space packed with electronics and a sparsely detailed chart just like ours.

"You are going north, up the coast?" he asked.

"Yes."

"Ah, good! Very beautiful. This"—he made a broad gesture out the windows at Nuuk's modest sprawl—"is not the real Greenland. But if you have a storm on your way"—he picked up a pencil and pointed at the tip of an unnamed peninsula—"here is good stay."

The chart showed nothing, not even the suggestion of a harbor or village.

"Ah!" he said, sensing my hesitation. "Yes . . ." He tore a page off of a pad and sketched . . . the end of a point, two big X's—presumably rocks—a dotted line drawn between.

"Here!" he said. "This way. Good stay. Safe trip!"

We shook hands and I walked back down the ladder, the paper with its precious advice folded in my pocket like a passport. We never visited this hidden harbor, but years later I would find the drawing, still tucked into a journal. Everyone we met extended this sort of generosity, typically a gift of food or some bit of shared local knowledge. A few days later at a fuel depot up the coast, I brought the gloomy superintendent a cup of coffee and he came alive, rolling out his own charts and pointing up a long blank inlet to a small crescent of harbor—his old village, now abandoned. "You must go," he said. And so we did, creeping up a calm bay dotted with icebergs until we inched past a curving sandbar and into the anchorage he'd described. There were still the bones of a few old houses along the shore. Next to one was an improbable spray of poppies, escaped from their abandoned flower bed and seeded downwind in a dazzling plume of color.

Just above the Arctic Circle is the departure point of Franklin's fleet, a broad coastal indentation set across the 70th parallel of latitude. Disko Bay is the size of Connecticut, sheltered by the dark bulk of its namesake barrier island and dotted with icebergs calved from the Jakobshavn Isbrae, largest and fastest among the dozens of glaciers that drain Greenland's ice sheet into the sea. Thirty-five billion tons of ice per year float away from the Jakobshavn glacier alone. I will zoom to the present for an example of just how much ice this is: When the container ship *Ever Given* famously ran aground in the Suez Canal in 2021, the flow of global commerce slowed to a crawl and the world awakened to the sheer size that such ships have attained. With her capacity of one hundred thousand tons, the *Ever Given* could haul away the yearly output of the Jakobshavn glacier in about 350,000 trips.

In summer Disko Bay sits for long periods under a calm dome of high pressure, jarringly serene in contrast to the punishment required to get there. Flat water, blue sky, the uncountable squadrons of icebergs like ghostly dirigibles floating past. Occasionally one will fracture and capsize with a roar of debris and a curling green wave, the fragments drifting away on the surge. At the highest tides, new bergs float forth from the glacier outlet in great abundance and the bay becomes all but unnavigable. Then one must wait, anchored in close behind a headland or tied up in the tiny harbor of nearby Ilulissat, cheek by jowl with everyone else who's had the same idea. Here at the wharf I watched growlers float around the harbor, ice chunks of a size that the International Ice Patrol helpfully likens to a grand piano. One had floated free on the tide and was bumping its way gently down the line of moored boats at the quay. In front of us a trawler crew, without much urgency, were pushing it away from their vessel with long poles, pausing every minute or so to take puffs on their cigarettes. Amid such a scene we were once visited by a group of American scientists passing through town after weeks spent inland doing fieldwork. We shared coffee while they told us of their project to collect core samples through the thickness of Greenland's massive interior ice sheet. They explained the records that were held by the ice, the chemical isotopes, trapped gases, and dust particles that would permit a time-stamped window into past climates— benchmarks for the broadening recognition that a man-made increase in global temperature was underway.

This was my first real glimpse at the working science of climate change, and the realization that the unexpected warmth of recent seasons at home could be part of something more global. I remember at the time feeling a bit incredulous. Just over the hill from town was the glacier—its jumble of frozen mountains grinding into the sea, their scale extraterrestrial. The idea that warming could gain a foothold in a place like this seemed to me abstract, if not far-fetched. One of our visitors zipped up his anorak and turned to go. The wind had picked up, I noticed, just enough to lift the earflaps on his hat. Standing unawares at the tipping point of an era, I bid him farewell.

"Good luck with your work," I recall telling him. "It still looks pretty cold here to me."

⌒

THE Northwest Passage is in truth really the Northwest *Passages*. There are in fact up to seven possible ways through the Arctic Archipelago, each with its own navigational challenges and contingent on the ever-present wild card of sea ice. A researcher named R. K. Headland at the Scott Polar Research Institute of Cambridge publishes a periodic total of all the ships that have so far traversed one of these routes successfully. The list is still brief. As of 2020, just 325 transits of the Northwest Passage had been accomplished since Amundsen first got through in 1906. You could watch this many ships pass in a day if you stopped to count by the Strait of Gibraltar.

The first navigation of the Northwest Passage by a merchant ship did not happen until 1969, when the discovery of oil on Alaska's North Slope led the Humble Oil Company (now Exxon) to send a modified supertanker called the *Manhattan* on a round trip from New York to Prudhoe Bay. The goal was to see if it made economic sense to bring Alaskan crude to market via a northern sea route. The short answer was no. There were diplomatic issues

connected to Canada's claim that the route lay within its territorial waters, and environmental questions over the risk of sailing
dangerous cargo through a pristine wilderness—but in the end
the job was just too damned difficult. On her outbound transit,
Manhattan faced the same challenges that had plagued everyone
from Franklin to Amundsen. She ran headlong into the torrent of
ice blocking the direct route west and was in the end forced to
divert into the tangle of geography nearer the mainland. The Canadian icebreaking ship *John A. Macdonald* followed like a sheep-
dog, eventually required to charge ahead and free the *Manhattan*
herself, held fast by a single pan of ice several miles across and
twenty feet thick. This was in late summer, with ice concentrations near their annual minimum. After another voyage in the
following April to test the ship in winter conditions the project
was shelved, and construction of the Alaska pipeline began soon
afterwards. Laying eight hundred miles of pipe across two Arctic
mountain ranges was deemed an easier solution for the transport
of oil than regular navigation of the Northwest Passage. Her year
of glory over, *Manhattan* resumed the prosaic life of an ordinary
tanker, hauling crude oil around the world's unfrozen oceans until she was driven aground in Japan by a typhoon and sold for
scrap in 1987.

According to Headland's list, *Manhattan* would be around the
tenth vessel overall to have traversed the passage, though she
turned around at Barrow, Alaska—short of the actual Pacific—
and is thus not technically eligible for the honor. In the post-
Manhattan years between 1970 and 1990 there were thirty more
trips altogether—mostly by research ships and icebreakers, but
also for the first time a passenger vessel and a few yachts, small
boats sailed by men and women who were taking on the Arctic
purely for the sport of it. During the 1990s there were an average
of three yearly transits, including appearances by the cruise ships

Hanseatic and *Frontier Spirit*—each a floating sign of things to come. Kevin Wood sailed the Northwest Passage himself aboard the US Coast Guard icebreaker *Healy* in 2003, on a voyage made without drama and with only moderate need for icebreaking. Despite the easy going, he was reluctant at the time to say that sea ice conditions had diminished beyond the natural realm of variability. After all, parts of the Arctic had warmed once before in the modern era, when a twenty-year period of diminished ice between 1920 and 1940 was followed by a return to much colder regimes.

In hindsight it's clear that something very different was afoot by 2003. Between 2001 and 2010 there were a whopping seventy voyages completed through the Arctic Archipelago, more than double the preceding ten years' total. In the year 2013 the bulk carrier *Nordic Orion* carried seventy thousand tons of coal from Vancouver over Canada to Finland, the first fully laden cargo ship to make the passage. Among working craft and lone adventurers now came the ships of the rich and famous, craft like the 46-meter *Dione Sky* and Paul Allen's mega-yacht *Octopus*, 128 meters in length. Between 2010 and 2020 were 180 vessels—including the thousand-passenger cruise ship *Crystal Serenity* and a veritable flotilla of yachts: again the mighty *Octopus*, as well as *Kamixtha*, *Infinity*, *Thor*, and my favorite, the *Upchuck*. If plotted on paper, a graph of ships per year would pivot between the millennia from flat to vertical. More vessels have navigated the Arctic Archipelago over the last decade than in all of previous history combined. The Northwest Passage, doom of so many, has in just twenty-five years become a route for commerce and adventure travel at the edge of a suddenly opened ocean.

Kevin Wood is the chief scientist of another project called Arctic Heat, measuring sea ice and energy flux in the Chukchi Sea, a marginal body of the Arctic Ocean bounded by the protruding masses of Alaska and Siberia. Broad and relatively shallow, the Chukchi is considered an important regulator of solar heat uptake in the Arctic, though its intermittent soup of broken ice makes it a hard place to gather data. The surface, too unstable for fixed instruments, is problematic for navigation and a meatgrinder for conventional buoys and sensor arrays. To solve this challenge, scientists make low overflights in a NOAA aircraft and drop a clever assortment of devices out through a pipe in its belly—including something called the Air-Launched Autonomous Micro Observer, or ALAMO. A wonder of the lithium-battery age, the ALAMO is a buoyancy-controlled float that takes sequential dives through the water column to profile temperature and salinity, reporting its data by satellite at each surfacing.

The findings of Arctic Heat support its own version of the trend spelled out in R. K. Headland's score sheet—which is that sea ice is vanishing faster now than at any other time in the era of scientific measurement. According to the National Snow and Ice Data Center, the total area of ice found in September of 2020 was 40 percent below its yearly average since 1980, a distributed loss of fifty thousand square miles per year. Imagine the annual subtraction of an Alabama-sized area from the American land mass. Even the ice that stays behind is not what it once was. Along with the reduced overall area, each year brings a smaller fraction of durable *multiyear* ice and a higher percentage of first-year ice, which is softer and more prone to melt as the summer days grow longer and warmer. The vanished pack leaves open seawater behind in its place, a surface nine time more efficient at absorbing solar energy. Bays and harbors freeze later in the year, or not at all. Less ice, more heat. More heat, less ice. There is more open water in the Arctic now than anyone has ever seen. With their winter highways in retreat, the dogsled drivers of the North are now keeping fewer animals. The canine population of Greenland is in a sharply curved decline, its graph reciprocal to rising ship traffic in the Northwest Passage. The 55,000 dogs of 1994 were by 2020 reduced to just 15,000.

Greenland's broad central plateau is five hundred miles across and two miles high, made mostly of glacier ice left from millennia of steadily compacting arctic snowfalls. The preferred mass unit of glaciologists is the *gigaton*, one billion tons. That is ten thousand *Ever Givens*. A 2019 study in the journal *Nature* looked at data from the entire ice sheet and concluded that Greenland has lost around four thousand gigatons of ice since 1992, at a rate that increased fivefold over the sampling period. Collectively the decade from 2010 to 2020 was a series of record years for glacial ice melt in the Arctic—with about three hundred billion tons vanishing into the sea each summer. Three million *Ever Givens*, fully loaded.

The last ten years have also set a record for ship traffic in the *Northeast* Passage—a waterway known more commonly as the Northern Sea Route—which connects Europe to the Pacific across the roof of Siberia. State-of-the-art tankers now carry cargos of liquified natural gas along this track from Africa to Asia, sailing in numbers that have more than tripled since 2018. Once unthinkable but now also beckoning is the all-out polar route, a beeline across the top of the world that is expected to be ice-free for large parts of the year by 2060. The national fleets of the Arctic countries stand at different stages of readiness to support their interests in this suddenly navigable polar ocean, where ice-capable military ships will be required to oversee an expanding stream of transients.

The Canadians, no strangers to ice navigation, are busy building an impressive new set of reinforced warships to patrol the Arctic Archipelago, a place whose territorial status remains unresolved with regard to international shipping. Is the Northwest

Passage more like the English Channel or the Panama Canal? That depends on who you ask. As with most qualifiers that involve cold and suffering, nobody can touch the Russians in the sport of icebreaking, a field they have dominated more or less since its inception. The Russians have upwards of forty icebreaking ships, including nuclear vessels that are spared the logistical challenge of refueling and boast an unlimited supply of hot water for their spa tubs. The United States military, not leading the pack, today owns as many polar icebreakers as it does sailing ships—that is to say, two. The venerable Coast Guard cutter *Polar Star*, launched in 1976, is expected to remain in commission until 2027, when it will be replaced by one of three new Polar Security Cutters—vessels that for now exist primarily on the lines of congressional budgets. In something of a paradigm shift, the new Polar Security Cutters will reportedly come equipped with Aegis Combat Systems—a weapons control suite first deployed for US Navy cruisers. *Polar Star* shares a berth in Seattle with the medium icebreaker *Healy*, which despite being medium managed in 2015 to be the first American surface ship to reach the north pole unassisted. Built primarily as a research vessel, the well-worn *Healy* was disabled by an engine room fire off Alaska in 2020, forcing the Coast Guard to cancel Arctic operations altogether for that year.

⌒

Above my bathroom sink is a photo of an outrigger canoe in the middle of a lagoon, heeled just perceptibly under the press of a single sail. The water is the color of a tarnished bell, reflected up to the overcast in a way that makes the whole image appear almost geological—like sedimentary strata, all in dully graduated hues taken from the same palette. Green water, pale green sky, the sand like stained bone. In the distance is a darkly blurred horizon—the lagoon's far shore, the tall palms just visible in silhouette. There is an ephemeral, not-quite developed feel to the image, like a fresh watercolor still sitting under its film of liquid. People imagine the Pacific in high-contrast vistas of verdant land and dazzling ocean, but in truth much of it looks just like this—damp and blurred and barely even there at all.

The place in this picture is Tabuaeran atoll, a tiny ring of land a thousand miles south of Hawai'i. Six miles across and seven feet above sea level, it will be obliterated by rising sea levels in a few human lifetimes—a bit of grit blinked out of the ocean's deepening watery eye. There are about two thousand people living in this improbable spot today, enjoying the remote abundance of a place where coconuts are free and the reef is still teeming with fish. When the water rises, they will all go live somewhere else.

Kiritimati, or Christmas Island, is 150 miles south of Tabuaeran, exactly at the halfway mark between Hawai'i and Tahiti. It is a much drier place, perched just beyond the steady rains of the intertropical convergence zone. When Captain Cook visited this scorched skillet of sand and scrub on December 24, 1777, it was empty, though archaeology shows that the Polynesians were once here too, perhaps pausing in mid-voyage to the greener shores of Hawai'i. Kiritimati is among the largest atolls in the world, a broad platform of coral perched atop the stone dome of a sunken volcano. There's normally not much dry land to an atoll, but here the lagoon has dried into a broad maze of salty ponds and flats of hardpan baked to concrete by the sun. Seen from space, the place bears a startling resemblance to a pork chop. At Southeast Point, the shank of the pork chop, the dry ground runs clear across the island, and tangled vegetation grows chest-high from horizon to horizon. It's as flat as Holland.

These atolls are neighbors in the Line Islands chain, among the most remote places where humans currently live. The first

modern settlers of the Line Islands were an assortment of plant-
ers and fishermen, followed by Allied garrisons during World
War II. In 1956, a new regiment of British engineers arrived on
Kiritimati and set busily about improving roads, airstrips, and
other bits of infrastructure, taking breaks from their work to watch
some of the largest hydrogen bomb explosions ever set off—final
firings in an insane fusillade that preceded the ban on atmospheric tests in 1963. Over time the foreigners departed and their
camps became villages, with names taken from the droll imagi-
nations of their imported occupants: London, Paris, Poland, and
Banana—tidy settlements of concrete block and palm thatch, set
against the shadow of a strange and terrible past.

At midday in London it is bone dry and white hot. I'm riding
to town with David Langston, the island doctor, perched on the
back of a motorbike he's shipped himself piece by piece in boxes
from New Zealand. David was born here, the son of a British
agronomist who came long ago and decided to stay. Over my
shoulder I can see the ship at anchor in the roadstead, a thin
thread of smoke trailing from her generator exhaust. We are the
first vessel to call here in three weeks. There are a lot of children
along the road, incongruous in neat English-style school uniforms as they pelt home through the sawgrass and palm thickets.

There are about five thousand people on Kiritimati now, who,
like the residents of Tabuaeran, are mostly transplants from the
Gilbert chain 2500 miles to the west. Widely spread as they may
be, the Gilbert and Line Islands are united provinces of Kiribati,
a young nation freed from Britain's Pacific territories in 1974.
Kiribati also includes the barely populated Phoenix Islands, last
known destination of Amelia Earhart. It is the only country on
Earth to occupy all four hemispheres. The national capital and
main settlements of Kiribati are on Tarawa atoll, a place best remembered abroad for a bloody battle between Japanese and

American marines during World War II. Tarawa today is the most densely settled island in the world, with about forty thousand inhabitants sharing just a few acres of ground. To alleviate this congestion the government has moved a steady stream of homesteaders to more spacious places like Kiritimati and Tabuaeran, where they build thatch houses and scratch gardens from the thin soil.

The national land area of Kiribati is about three hundred square miles, spread out across 3.5 million square miles of ocean. That's a land-to-water ratio of 0.00009. The thirty-three islands have an average elevation of about five feet, rendering them an unhappy bellwether for sea level rise. If you piled them all up like pancakes, you'd wind up with a stack only two hundred feet high. In the winters of 2014 and 2015, flooding overran Tarawa. The water ruined gardens, swept away houses, and forced an evacuation of the city hospital. Astronomical high tides and an El Niño event were partly to blame, but an increase in sea level heights worldwide had reduced Tarawa's margin of dry land to zero.

Henry Genthe is an American who lives on Kiritimati, and as we drive around together in the backcountry, clouds of nesting seabirds rise up around us like windblown leaves. It is getting harder to dig a reliable well here, Henry tells me—no doubt in part because more people are using water, but also due to what he thinks is a changing hydrostatic balance between the surrounding ocean and the thin lens of fresh water that sustains life on the island. Coconuts, the sole cash crop, are suffering under the dual pressures of overharvesting and an increase in salt spray.

As with the old story of frogs in a kettle, the locals seem to lack a shared sense that a problem is at hand.

David Langston's father, Percy, is a slender Englishman wearing square spectacles and baggy shorts, a servant of the Queen gone permanently afield. He is a nonbeliever when it comes to sea level rise.

"The island is not sinking," he says. "Sea levels are not rising. See this lagoon here? We used to catch mullet in there and that's all dry land now. If anything, the area of the island is increasing."

Claims of sea level rise, Percy contests, are born here not from fact but a desire for assistance funds in a place perpetually starved of resources. There are a lot of churches on Kiritimati, and their influence on public opinion is also evident.

"God will decide," says someone, "whether these islands stay or go."

IN 2014 Kiribati's president, Anote Tong, forced into a visionary role by geography, spent $7 million on a large piece of ground in mountainous Fiji for his citizens to inhabit once their atolls are gone. Called Natoavatu, it is a 5500-acre freehold on the island of Vanua Levu, purchased from the Anglican church. A mixture of fields, hilly forest, and mangrove swamps, it currently is inhabited by descendants of another refugee population: Solomon Islanders brought to Fiji as slaves in the late nineteenth century. Tong's bold plans have since been shelved by the subsequent administration, in a swing toward the strident denial so typical of our times. When last I looked, there was news of a pending venture to develop some of the land at Natoavatu for agriculture and other unspecified uses. There is no legislation yet in progress—in either Fiji or Kiribati—to govern just when or how a formal emigration might be managed.

Meanwhile the water continues to rise. NOAA estimates that average global sea levels increased by about 3.5 inches over the twenty-five years preceding 2020. That's an eighth of an inch per year, which seems small until you do some math and see that's just over a foot per century. To paraphrase a popular expression, a little bit here and a little bit there, and pretty soon you are talking about real money. The average surface temperature on Earth rose about 1.2 degrees Celsius over the interval between 1880 and 2020. Most of the increase, a full degree, is crowded into the period since 1950. Ocean temperatures are rising too, though more slowly, due to the immense heat capacity of water. Nonetheless, the overall amount of heat stored in the oceans has grown enormously. Scientists estimate that the oceans have collectively been absorbing the equivalent of one or two atomic bombs *per second* of extra energy from the warming Earth over the last seventy years. Rather like air, water takes up more space when it is warm, and since the oceans are hemmed at their edges by continental shorelines, the only place the warmer water has to go is up. This is called *thermosteric rise* and explains about 30 percent of the change in sea levels that warming has induced. The rest is down to simple melting, as glacier ice turns liquid and runs into the sea.

Far in the hinterland of Christmas Island, there is still plenty of dry ground, for now. We've been driving all morning, circling the paved ring road and exploring sidetracks where salt pans shimmer in the heat. Henry shows us Aeon Field, one of three runways built for military aircraft that now languish under the sun, halfway to the southeast tip of the island. It's a two-mile strip, engineered to handle Vulcan bombers with nuclear payloads. Today the edges are crowded with shrieking boobies and creeping vegetation, but the center is still clean and empty, an invitation to drive at imprudent speeds as the faded white lines fly by.

Down the road we slow to a safer pace and stop for a break. A few of our group wander off to photograph birds, and I flip through my notebook while Henry dozes with his feet up on the passenger door. I hear a buzz in the distance, a noise that soon resolves into a moped loaded down with two fishermen and their gear, weaving sedately into view up the road. I spot the silvery tail of a large fish protruding off at an angle from an open backpack. The road is wide, but they manage in slow motion to collide perfectly with the open door of the truck before careening off into the underbrush. No one is hurt.

Henry starts briefly awake. He touches his hat with a nod of recognition.

"Gentlemen," he says.

They regroup wordlessly and ride away.

~~~

GEOLOGICAL records show that our climate has been through many long cycles of warming and cooling, spread out over the eons since Earth moved past being just a big smoking ball of iron. This evolution includes a repeating pattern of ice ages, periods when the poles are covered by ice and long cold spells alternate with short warmer intervals. The earliest identifiable ice age was the Huronian, which happened around two billion years ago at about the time that oxygen first became common in the atmosphere. There are suggestions, albeit questionable, that ice sheets once reached all the way into the tropics during a so-called Cryogenian, or "Snowball Earth" period. Measured in geological time, ice ages are brief events, a few million years of cold separated by much longer spells of uniformly mild conditions. Believe it or not, we're technically in an ice age right now called the Quaternary period, which began about 2.5 million years ago. Climate records show a trend within the Quaternary period of warmer and colder spells, with times of extended glaciation separated by warmer interglacial periods of about twenty thousand years. We're enjoying one of these warm periods at present—the Holocene epoch— which began around ten thousand years ago, roughly even with the emergence of agriculture and long-term human settlements.

During an even warmer stretch 130,000 years ago called the Eemian period, there were hippos wading in the rivers of Europe. This ended when glaciers pushed south again, back onto the Asian and North American continents. The hippos moved on, and things grew sharply colder. The ice came as far south as Cincinnati. It left gouges in the rocks of New York's Central Park and plowed a wayward tongue of gravel into the hills of Cape Cod and Nantucket. Broad areas of prairie and forest gave habitat to mastodons, giant sloths, and saber-toothed tigers. There were people walking around at this point too, but only a few. Estimates are that most of Earth's human population in the year 100,000 BC would have fit into a large football stadium. At the start of the Quaternary glaciation, 2.5 million years ago, early hominids were just starting to appear in present-day Africa. It would be another two million years before they used fire to cook.

Glaciers are built by the year-to-year accumulation of snow that has survived the attacks of melting, wind erosion, and *sublimation*, the direct conversion of ice into water vapor. Over time, the jumble of ice crystals gets denser as gravity compresses the voids in between. Tiny bubbles of atmosphere stay trapped in the grains, so that if you put a chunk of glacier ice into a glass of water, it will fizz like soda. In warm spells, the snow is not replaced fast enough to match the rate of depletion, and glaciers retreat. During prolonged periods of cold, glaciers grow and flow outward like very slow-moving rivers. They grind down valleys and across landscapes, doing considerable damage on their crawling way. Glaciers dug the Great Lakes, built the fjords of Norway, and left behind the *erratics*—garage-sized rocks that dot northern woodscapes like grounded asteroids.

The intellectual concept of ice ages emerged in the nineteenth century, as scientists in Europe wrestled with the recognition that historic forces had been at work moving the big rocks around in their respective mountain valleys. They blamed the glaciers. Large ice masses sat heavily among the high peaks and cols of the Alps, and from rock samples it became evident that these cryonic scouring pads had made multiple passes across the landscape over time. The term *ice age* is itself credited to a German botanist named Karl Friedrich Schimper, who introduced the title in 1837 to describe the concept of periodic glaciation. Schimper's work ran parallel with that of several contemporaries, among them the Dane Jens Esmark, who was first to suggest there might be a connection between climate change and variations in Earth's orbit. By the late 1800s, enough evidence had accumulated to support a move away from the old model of a steadily cooling Earth to a new theory of change between alternating periods of warm and cold.

Identifying the cause and the timing of these changes would leave plenty for successive generations of scientists to work on. A Croatian engineer named Milutin Milanković spent considerable time in the early twentieth century on models to project the correlation between ice ages and three known variables in Earth's relationship to the sun: our changing elliptical orbit, a long wobble in the planet's rotation, and the periodic alteration of *obliquity*, which is the relative angle between Earth's axis and the plane of the solar system. A specialist in the new field of reinforced concrete, Milanković seems at first an unlikely father to the founding canons of climate science—but this was the age of Einstein, a time when patent clerks came home from their day jobs and set about unlocking the secrets of the universe with pencil and paper. Among his accomplishments, Milanković was able to make predictions about the climate on the moon. It would be another fifty years before anyone went there to check, but time has proven his predictions to be startlingly in the ballpark.

⌒

THE scientists who visited us in Greenland were part of something called GISP-2 (Greenland Ice Sheet Project 2), a multiorganizational effort that would ultimately collect a continuous

110,000-year record of past climate preserved in ice cores. To a glaciologist, an ice core is like a tree stump, with each annual layer revealing its own bit of the past. The Greenland ice sheet is thicker than any tree, and the first challenge for the GISP-2 team was the accurate identification of each layer, all 110,000 of them. One for each year. This meant more than simply counting rings, and in the end required an array of measurements to be sure of when one year ended and the next began. Each layer held a trapped atmospheric record for the year it was deposited, and once the time scale was established, the researchers set about measuring how things had changed along the way. Concentrations of sea salt and dust served as proxies for wind conditions in a given period—in Greenland, salt reaches the ice cap on southerly winds from the Atlantic or on easterlies driven by a recurrent low-pressure regime near Iceland. Calcium-rich terrigenous dust (a.k.a. dirt) is borne mostly by westerly continental winds. In colder periods a strong polar jet stream produces stormier conditions at the surface, with shorter summers and less plant growth to protect land areas from erosion.

For the scientists of GISP-2, year-to-year levels of gases and aerosols were measurable in the ice, either as solutes in the water itself or trapped as gas bubbles in each layer of core. The resolution was clear enough to identify fallout from postwar nuclear tests and the Chernobyl reactor accident in 1986. It was a heroically tedious task, much of it conducted on scene by people clad in freezer suits and polar gear to fight hypothermia. The reward was a timeline of atmospheric data more than a thousand centuries long, with a previously unseen level of detail.

Ice records like the GISP-2 cores have gone far in confirming a connection between the Milanković cycles and climate change, though the many long warm epochs across history show that solar cycles alone are not enough to guarantee regular ice ages.

After all, dinosaurs walked the planet in toasty contentment for a hundred million years. There might even have been palm trees at the north pole. It turns out that other influences are also acting at the control panel of climate. Driven by the process of plate tectonics, the continents are adrift on Earth's surface, moving just about as fast as your fingernails grow. Since ocean currents are the main mover of heat between latitudes, the location of land will have a lot to do with where it's warm and where it isn't. Today's arrangement—with a closed Arctic Ocean and large north-south continental formations—is quite limiting to heat exchange. The start of the geologically recent Quaternary glaciation coincided with the closing of the Panamanian isthmus, which effectively shut the door on the movement of warm water between Atlantic and Pacific. Much longer ago, the warm, ice-free Mesozoic era began with most of Earth's terrain packed into a single giant land mass called Pangaea—an arrangement that left the remaining oceans free to circulate heat uniformly. Pangaea itself was warm and arid, particularly in the vast hinterlands that were insulated from oceanic moisture—think Australia. Things changed at the beginning of the Jurassic period, as the supercontinent fractured and new coastlines gave inroads to moist ocean air. These dinosaur boom years were still warm, but much rainier as wet maritime air found new paths ashore.

As recent history has taught most of us, the chemical balance of the atmosphere is another critical mediator of climate. Heat is held in the atmosphere by greenhouse gases, whose molecules absorb and delay outgoing infrared radiation on its path back into space. The greenhouse gases collectively keep Earth at an average temperature of around 60 degrees Fahrenheit—a good deal warmer than it would be otherwise. This so-called greenhouse effect is rightly listed among our current pressing issues, but it's important to remember that in its total absence the mean temperature on our planet would hover near minus 10 degrees Fahrenheit—too cold for liquid water, the key ingredient of life as we know it. The principal greenhouse gases are carbon dioxide, methane, nitrous oxide, and water vapor. Despite all that we hear about them, they are pretty scarce. CO_2 exists today in concentrations of about 420 parts per million (that's 0.042 percent) and was made mostly by volcanoes until humans began burning carbon fuels several centuries ago. For an example of what too much carbon dioxide can do, one need only look to Venus—where the atmosphere is 97 percent CO_2, and the surface temperature is 872 degrees Fahrenheit.

CO_2 levels in Earth's atmosphere have gained worldwide attention since a scientist named Charles Keeling began a series of measurements atop the Hawaiian volcano Mauna Loa in 1957. Keeling set up his instruments and almost immediately noticed a couple of very interesting trends: First, the graph was jiggly, bouncing up and down like handwriting done on a city bus. The jiggles had a semiannual period, and Keeling et al. quickly realized that they were the product of photosynthesis. Each spring the trees of the northern hemisphere were taking a collective gulp of CO_2, only to drop their leaves and exhale it in the fall. Adding data from paleoclimate studies, Keeling and his team extended their plot backwards into time and saw something that concerned them. Drawn against the arc of history, the curve of CO_2 concentrations remained relatively flat until the mid-nineteenth century, the point at which consumption of fossil fuels went into overdrive to supply power for industrialization.

From there the curve looked suspiciously exponential, showing the distinctive hockey-stick profile that a value takes on when it is being multiplied at regular intervals by itself. In 1850, CO_2 comprised about 260 parts per million in the atmosphere. In 1900, the fraction was about 300 parts per million. In 1957, the start of the Mauna Loa measurements, it was 340 parts per million. Given the evidence of how CO_2 concentrations had affected earlier climates, this was a disquieting trend. By early 2021, the number was bumping up against 420, a 60 percent increase from its preindustrial value. All this extra greenhouse gas is credited with forcing the global temperature rise, estimated at about 1.2 degrees Celsius, that has taken place over the industrial era—all while the continents have stayed put and variations from the Milanković inputs have remained fairly stable.

Professor Paul Mayewski is director of the University of Maine Climate Change Institute and was a leader of the original GISP-2 project in Greenland. At a 2021 Earth Day presentation to the Maine Sierra Club he showed a data graph to the audience, a long serrated line hovering close to the X-axis and then bolting skyward at the right-hand margin of the screen. Imagine the heart rate of a ship's captain, sleeping peacefully and then suddenly awakened by a panicked shout from the watch. The line is Keeling's curve, with an 850,000-year interval of ice core data hitched to its tail. From this figure it's plain to see that atmospheric CO_2 levels are greater now than they have been at any time in nearly a million years. It's not even close. At the rate we're going, they will be near 520 parts per million by 2060, a doubling in just over two centuries.

The global change in temperature from greenhouse warming has been more visible in some places than others. Around certain parts of Antarctica things are actually a bit colder, as stormy conditions in the Southern Ocean have forced the upwelling of cold seawater along the continent. Just about everywhere else it is warmer, in some cases a lot warmer. Climate scientists like to speak in terms of *anomalies*, the deviations from a standard mean over time. Color-coded maps are a good tool for this, and Dr. Mayewski displayed a world map where decadal temperature variations are shown in relative hues of blue (colder) and red (warmer). Nearly the whole planet is painted somewhere in the red spectrum—most of all the eastern Arctic, where mean temperatures have risen as much as 5 degrees Celsius since 2004. A great red blob is perched somewhere north of Siberia, in a big empty space that would once have been ice but is now open ocean for large parts of the year.

The recent progress of science indicates that we are not the first population to wake up and find one set of years abruptly quite different from the last. Sudden past swings in climate are in fact robustly evident in the geological record. The GISP-2 cores are no longer new, but the imaging of preserved samples with modern instruments has allowed what Mayewski calls a "storm-scale" resolution of certain events, a glaciological version of what Old Weather has done with ships' logs. By this new metric, the thermal jump from the cold Younger Dryas into the warm Holo-cene epoch is visible across a span of mere years, not decades or centuries as was once thought. Preserved in ice, the Holocene transition inhabits just a short core section from about a mile down the GISP-2 borehole—a meter of material representing a thirty-two-year period between 11,643 and 11,675 years ago. In the data, seasonal evidence of airborne salt and calcium—glaciological proxies for marine and continental winds—tails off sharply after a short interval of particular turbulence. It is a bold marker for a new period of warming, as a poleward retreat of the jet stream led to longer and milder summers in the Arctic. In a 2014 paper authored by Mayewski and his team, these findings read like a real-time weather synopsis, broadcast from the age of mastodons: Three years of rising westerly wind peak values, a transitional year of retreating sea ice and increased maritime airflow, then an abrupt relaxation into the longer and more placid summers of the Holocene.

I imagine this remarkable summation as it might play out in the laconic discourse of waterfront communities:

"It blew hard the last few years, but things have eased off in a hurry, haven't they?"

"Yup, with the ice melting like it is you'd best move your boats and houses up the shore soon."

From pondering transient boulders in mountain passes, climate science has come to the point of delivering date-stamped characterizations of weather from 12,000 years ago—cautionary tales from a time when the speed of events seems to mirror the urgency of our own circumstances.

To Kevin Wood, one challenge of basing climatology on historical accounts is an understanding that early measurements were not standardized and no means existed to image large areas in real time. Who could really be sure what the big picture was on any given day in 1854? This was a particular challenge in polar regions, where ice conditions could change abruptly in response to wind shifts, leaving whole fleets trapped or suddenly free—floating in broad open reaches with no ice in sight, and no way to tell where it had gone. Today, the old baselines are changing so quickly as to not be baselines at all—each year a record for warming or nearly so, the annual minima of ice in steady retreat, and the seasonal boundaries of weather events utterly disrupted. Chicago freezes. Siberia bakes. The arc of climate crumbles at both ends into uncertainty. Arrayed in my kitchen each spring are green flats of nursery seedlings, living in sheltered warmth until May 15, the official threshold for when it's safe to set plants outside in Maine. No prudent gardener would violate this edict, any more than a sailor would set out for the West Indies before late

October—when the summer ocean starts to cool and the prime months for hurricanes are safely past. Or so the assumption has been until recently.

In September of 2017 Professor Kerry Emanuel gave a talk at MIT about the effects of global warming on the frequency and severity of tropical cyclones, beginning his presentation only hours after Hurricane Maria began cutting its awful swath across Puerto Rico. This was in Dr. Emanuel's own words a tragic irony, as he had composed his lecture about the year's unprecedented hurricane season well before Maria exploded into a category 5 storm just east of the Windward Islands. The year 2017 was off the charts for tropical cyclones in the Atlantic, the prize winner in one grim category after another. Of the seventeen named storms, ten were full-fledged hurricanes, with sustained winds above 64 knots. Two made landfall as category 5 systems, with winds above 140 knots. There were $300 billion in damages and over three thousand fatalities. On August 24, Hurricane Harvey came ashore as a category 4 storm at San José Island, Texas—the first of three landfalls Harvey would make as it bounced sluggishly up the Gulf Coast, dissipating its vast energy in a biblical deluge. The city of Houston received forty inches of rain in a four-day period. There were 106 deaths and $125 billion in property losses—making Harvey comparable in scope to Hurricane Katrina twelve years previously, among the costliest natural disasters in American history.

Hurricane Irma formed just west of the Cape Verde Islands on August 30, right about the time that Harvey was dissipating into a loose mass of thunderstorms over the American Southeast. I remember gaping at satellite images of Irma making its way across the Atlantic, a tightly coiled and beautifully symmetrical spiral of clouds. On September 6, Irma's winds peaked at 155 knots, making it the strongest cyclone of the year worldwide. Irma flattened Barbuda, St. Martin, and the Virgin Islands before pummeling Cuba and just missing Key West on its way to bludgeon the Florida panhandle.

Maria developed just east of the Lesser Antilles on September 16, deepening rapidly and destroying the tiny island nation of Dominica two days later. Maria supplanted Irma as the year's most powerful storm and went on to make a direct hit on Puerto Rico, where close to three thousand people died. Parts of the island remained without power and water two years later. Harvey, Irma, and Maria were numbers three, four, and eight in a ten-member string of consecutive hurricanes in 2017—a streak unduplicated in the era of satellite observation.

A large part of Kerry Emanuel's work is to determine whether the recent surge in deadly tropical weather events is part of natural variability or coupled to man-made changes in climate. He is well convinced of the latter after comparing historic data with the results from modeling experiments. The modern increase in greenhouse gases means that the oceans are staying warmer, and as a result sending more water vapor into the atmosphere to power storms. Thanks to something called the Clausius-Clapeyron equation (says Professor Emanuel in the casual way that scientists have), we can predict that a one-degree increase in sea temperatures will translate into roughly a 7 percent uptick in airborne water vapor over the oceans. Thermodynamically, the potential maximum speed of hurricane winds has increased by about 6 knots per decade since 1980. This all translates into bigger, wetter storms, something that the record for the last forty years appears to bear out. Hurricanes today are reaching their peak intensity at higher latitudes, carrying energy farther away from where we have customarily considered the tropics to be. Using values projected for a virtual century beginning in the year 1990, the likely chance of a

Harvey-strength storm impacting the state of Texas will undergo a twentyfold increase in probability—from a hundred-year event in 1990 to a five-year event by 2090.

The other thing that Kerry Emanuel wants us to understand is that the most dangerous part of hurricanes is really not the wind; it's the water. Sailors in boats tend not to think this way, but he's got the data to prove it. Of the 589 Americans killed by hurricanes between 1970 and 1999, 85 percent were drowned. Flooding in Honduras from Hurricane Mitch claimed as many as ten thousand victims in 1998. Nobody sends a TV reporter to stand in a puddle while they report on a hurricane, says Dr. Emanuel. Windblown palm trees are much more photogenic, but it is the water that is lethal. The people of New York and New Jersey would realize this in 2021 when the attenuated remnant of Hurricane Ida made its way up the Appalachians after crashing into New Orleans on August 29. Packing 150-knot winds on landfall, Ida matched Hurricane Laura from the preceding year as the strongest storm to ever hit Louisiana, but its cost to human life was greater in metropolitan New York, where seven inches of rain fell on Central Park in one day. In the borough of Queens people drowned in their basements. Cars were swept away by flash flooding in places like Hoboken and Mamaroneck, a leafy suburb where I remember visiting relatives with my family, long ago. An earlier, less potent Hurricane Ida came ashore in 2009 at Mobile Bay, Alabama—startlingly close to the landfall of its 2021 namesake when one considers all the places that a tropical cyclone might go. Probably now the name will be retired.

Along with the rain, nearshore communities must also be prepared for storm surge, something that Professor Emanuel likens to a wind-driven tsunami. Storm surge can take the form of a very rapidly rising tide—as when Hurricane Sandy inundated the subway tunnels of Manhattan—or can come hurtling in all at once like the outflow from a breached dam. There is a shocking video of this now popular with lecturers, taken in the Philippines during Typhoon Yolanda in 2013. A raging wall of water comes out of nowhere to obliterate a row of wooden structures along a stricken waterfront, a moment caught on film by an individual who somehow survived the experience. Surge for specific storms is affected by the relative stage of the tide. As a child, I was stranded with my mother in a Boston hotel during the February blizzard of 1978. We were there for close to a week, in the end letting ourselves into the kitchen to help the few staff who were still in the building. During this hurricane-strength winter storm the surge arrived together with a full-moon high tide to create floods fifteen feet above normal in some coastal areas. Breaking waves made houses vanish and washed loaded trucks into the ocean.

This sort of thing is not a new hazard, but it is an urgent one, as coastal reinforcements are made obsolete by rising sea levels and the energies of global warming turn hundred-year storms into decadal occurrences. Nearshore populations in North America have tripled since 1970. We're not ready for this, Dr. Emanuel reminds us. There are too many arrows pointed in the wrong direction. Near the end of his talk he flashes a colored map of some unnamed estuary, bounded by a long peninsula dotted with tidal lagoons. It is Cambridge, Massachusetts, under the projected conditions of a three-foot storm surge modeled for the year 2070. Much of the MIT campus is painted in blue, unmistakably underwater.

In September of 2020 I am standing in the main classroom building at Maine Maritime Academy, a small public university where ambitious young people rise early each day to train for careers at sea. I have taught here on and off over my own career, and today my former students are docking ships in Port Everglades, farming oysters in Maryland, and drilling for oil off of Africa. The academy is a transitional world—part ship, part college, where some of the professors are called "Captain" and the polished hallways are set about with bits of nautical memorabilia. Near some framed portraits is a video screen that shows the current North Atlantic weather map, an ever-changing mural where today in the tropics there are no fewer than five active cyclones. Hurricane Sally is pounding the Florida panhandle while Hurricane Teddy aims ominously toward Bermuda. Vicky, just named, is west of Africa—near another embryonic knot of clouds which, when fledged, will be called Wilfred. At this point the National Hurricane Center will be out of proper names and obliged to apply Greek letters—starting with Alpha for an unusual storm west of Lisbon and Beta for another nascent system hovering in the warm Bay of Campeche. It will be apparent in the weeks to come that we are nowhere near finished. A parade of cyclones awaits, deadly storms with names like Gamma, Delta, Epsilon, Zeta, Eta, Theta and Iota. Iota will strike Nicaragua as a category 5 hurricane, only two weeks after a direct hit from Hurricane Eta. Never have so many tropical cyclones been recorded together, and not since 2005—the epic year of Emily, Katrina, Rita, Wilma, Alpha, Beta, Gamma, Delta, Epsilon, and Zeta—have we run short of real names to call them by.

Do sailors now see the effects of climate change at sea? Are the storms worse? For working mariners the one-storm-at-a-time imperative of seafaring can make longer patterns more difficult to visualize. But here we are, I think, in mid-September, with two months left of hurricane season and already out of names. This once-in-a-lifetime occurrence has happened twice since I bought the car I am driving. It will happen again in 2021. On the radio I hear that America's dry western forests are ablaze from Los

Angeles to Seattle. The jet stream carries their microscopic particles to us here in New England, a high plume of dust that's drawn a pink-gray scrim across the early-morning sky. As I drive, the road cuts abruptly to the east and the sun hangs briefly, tallow-colored, at the center of my windshield. It is sharp-edged and corpulent, seemingly near enough to visit. There is a star out there, I remember—a thermonuclear forge. It is ninety-three million miles away but, like an overfilled woodstove, suddenly too close for comfort.

12

HTHH

Through the ship's fitful internet connection I am reading email from home. I see that my electric bill is due, and that a fight is brewing over some new construction at the boatyard up the street from my father's house. How happy to be here. Across from us in Pago Pago's remarkable harbor are the cannery wharves, a seaside industrial park with row upon row of hulking tuna seiners rafted alongside—space-age fish-killing machines with an architecture surprisingly graceful for vessels of modern commerce. The long low ships and humming factories are like a secret city, its workings opaque and inhabitants nearly invisible to the surrounding world. Occasionally I meet the boat crews here in the village—slender Asian men in sandals and windbreakers, buying cigarettes or solemnly counting stacks of currency to send home at the Western Union office.

Up the road, past McDonald's and the town square, are the cargo wharves—their fenced boundary just opposite a row of

trees filled with what first appear to be crows but are in fact speci-
mens of *P. samoensis,* a disconcertingly large bat that will soon be
featured on a Samoan version of the American twenty-five-cent
coin. A family in a pickup truck brakes to let me cross the road,
and I ask a guard near the terminal gate for directions to the port
offices. I've come to inquire about the arrival of a wayward shipping container, now weeks overdue. Anxiously awaited, it has become snagged like a lost package in the tortured logistics joining
mainland America to a territory that few even realize exists. Our
lost box began its journey months ago in Massachusetts, packed
with the vital bric-a-brac demanded by a ship operating in distant
waters: Peanut butter, microscope slides, engine spares, power
strips, and monitors. Marshmallows. Cases of canned fruit, heavy
as uranium ingots. Also twenty reams of nonmetric printer paper,
a half mile of rope, forty-one pillows, three life rafts, a small re-
frigerator, and a new mainsail. Such are the contents of these steel
crates that appear in all corners of the world, anonymous vessels
for anything that will fit into them. Running shoes, cheap furniture, tractor parts, bagged cement, sewing machines—or a Lam-
borghini. Once, long ago in the Mediterranean, a friend of mine
watched a container fall from a ship's crane, bursting open to reveal a lifetime supply of shish kebab skewers.

A container bound for American Samoa from the mainland
begins its journey with a truck ride to the port of Los Angeles—a
veritable city of containers—where it will wait among a hundred
thousand cousins until it is loaded onto a ship for Hawai'i. Five
days later it will arrive in Honolulu, resting again until space is
available aboard another ship making the far less frequent run to
Pago Pago. Thanks to an obscure century-old statute called the
Jones Act, only American-flag vessels may carry cargo from one
US port to another, thereby granting a monopoly to the few

native companies whose ships still sail between America and her Pacific outposts.

Due to a clerical error, our particular box has, literally, missed its intended boat. A cascade of systemic disruptions is also at play, triggered back in February when a key ship in the trading fleet cracked open from sheer antiquity and began leaking oil into San Francisco Bay. The SS *Matsonia* is a *con-ro* (vehicle and container) ship from the obsolete Ponce de Leon class, last surviving sister to the lost *El Faro*. She is in her final year of scheduled service, but her unplanned visit to the shipyard has thrown ripples across the whole pond of South Seas logistics—raising doubt as to whether or when our expensive care package will reach us at all. Our agent chooses optimism. Next Saturday for sure, she promises—though the port is closed over the weekend, so . . . Monday.

It is October, the turning point of what will be spring in the southern hemisphere, and a deep meander in the trade winds is pushing a gusty band of rain toward us from the east. There are gale warnings posted for tonight, the long funnel of Pago Pago's harbor forcing a lumpy sea right up to our dock while the ship surges like a leashed Airedale and the crew struggles to keep the fenders in place. We are floating in the flooded maw of an extinct volcano, its steep green hills rising up in every direction and blurring invisibly into the overcast. Some of the crew are headed off for lunch, and I join them in our car for a drive out the island's only road. The speed limit is 30 miles per hour everywhere in American Samoa, without much apparent disagreement. Why hurry, in a place with so little driving to do?

As we ride by the harbor entrance, I am grateful we are not leaving today. The rain pelts down and rollers break onto a broad shallow area just west of the main harbor approach, a broad gray maelstrom of fractured ocean. The entrance channel to Pago Pago is called Narragansett Passage, its decidedly un-Samoan name taken from a ship that brought representatives of the US Navy here to make trouble in 1872. One of the oldest Polynesian kingdoms, Samoa enjoyed 2500 years of unmolested autonomy until its convenient location in the middle of the world's largest ocean caught the interest of naval powers in the nineteenth century—all of whom coveted such places as coaling stations for their steamships. Things came to a head in 1889, when a standoff between German, British, and American warships was interrupted by the unexpected arrival of a hurricane. With weather's typical impartiality, the storm destroyed all three fleets and moved on, effectively deferring the resolution of colonial dominion for another generation. The western parts of Samoa were occupied by Germany until World War I and went on as a territory of New Zealand until their independence in 1962. The eastern island of Tutuila and a few others nearby have stayed American, evolving from a military enclave to a self-governing territory in the mid-twentieth century. American Samoa is the only piece of the United States below the equator, save for a tiny uninhabited rock called Jarvis Island about 1500 miles south of Hawai'i.

On our way back to town we pass through gardens of breadfruit trees and banana fronds nodding in the wind, with tall rows of coconut palms behind them in a tilting overstory. Outside a small village we pull over and wait for *Sa*, the traditional prayer curfew that's practiced here each evening. A large man wearing a lavalava and an Oakland Raiders jersey walks past us and strikes a gong made from an old propane cylinder hanging in a wooden frame. He nods hello on his way by. Today the running of things in both halves of Samoa is left mostly to the Samoans, who are no doubt happy with that. The American and independent parts are linked by a shared language and genealogy but now bifurcated by geopolitics. Young men from American Samoa, someone tells me, have the highest rate of enlistment in the US military and are 175 times more likely to play football in the NFL than members of any other population. The independent western islands remain in the cultural orbit of Fiji and New Zealand and are thus wild for rugby. I suspect I am receiving an abridged narrative reserved for visitors—three thousand years of history reduced to pointy balls and blindside tackles.

〜

WHEN our container finally arrives, it is like Christmas, only better, a great gift box with its doors thrown open and the crew rooting gleefully for long-awaited supplies. I think again of rugby. At the center of the scrum is Penikolo, a young scientist from the Tongan Department of Environment who's joined us for our upcoming departure. He is friendly but with a stranger's reflexive reserve—a trait he balances by being what seems like everywhere at once, somehow always there to carry the other end of the object you have just lifted. The weather is better now. Tomorrow we are bound south for Tonga and then to Fiji, each port a step in our long navigational arc down to Auckland. It is a path engineered to remove us from the tropics before the onset of cyclone season. I picture the vast nearby sink of the Western Pacific Warm Pool, and think again about the colonial warships that were here just over a century ago—itching to fight and oblivious to the danger awaiting them from larger and less partisan forces.

We are in the all-or-nothing time of predeparture, the final details crowding together like commuters at a turnstile. The crew is busy with our new mainsail—big as a circus tent, cascading in folds over the quarterdeck as it is bent on and lifted slowly onto the boom. Penikolo and two helpers are unpacking box after box of stores, and I notice that there is an air compressor running on the dock, next to a snarl of black hose leading down into the harbor. At the surface a moving column of bubbles discharges it-self along our waterline like bursts from a soda siphon.

Before we are allowed in New Zealand, we must first have our bottom cleaned, ships being as they are a stealthy accomplice to the great scourge of invasive species. Hitchhiking fish swim hap-pily in the seawater that many vessels carry as ballast, while a host of less charismatic biota cling to hulls and propellers. Since 2004, ships have been required to pump their ballast overboard at each international boundary and exchange it for local water—thus pre-venting a French eel, for example, from sneaking a ride to raise its unruly children in the unsuspecting harbors of Borneo. The In-ternational Maritime Organization has now turned its attention to the more sessile realm of barnacles, worms, and algae—perhaps undifferentiable to the novice eye, but still a risk to ecosystems if loosed in the wrong location. Justifiably protective of their island paradise, the Kiwis are early adopters of the new standards, and in their chipper way assert that all foreign ships must now be cleaned of growth—with their bottoms scrubbed or painted—no more than thirty days previous to their arrival.

A loop of hose emerges from the harbor like a fishing line cast in reverse. At the end of it is a young woman in a wetsuit, her wet cloud of hair bisected by the strap of a diving mask. She climbs out of the water on an aluminum ladder that's been leaned against the wharf, a pirate princess emerging from some hidden aquatic realm. Arresting, if not gregarious, she lifts her toolbox and wres-tles the hose into a pickup with practiced flips before turning to observe me and speak for the first time.

"All done, Cap," she says. "My father will be back later with the videos. And the bill."

Underwater videos, the final proof that we have purged our bottom of bad actors from the lower branches of zoology.

"Thanks for helping us out with this," I say. "I know you guys are busy."

The pirate reaches for her cigarettes. "Yup," she says.

She wheels away in her truck, a tycoon in this singular trade. I imagine a chalet in Banff, or a perhaps just a row of oil drums in a boat shed, filled with hundred-dollar bills.

"How do people know I am the captain," I ask my wife, "when I haven't even introduced myself?"

"That's easy," she says. "It is the way that captains stand around. Like they are doing nothing yet somehow busier than anyone else."

ONGA is three hundred miles south of Samoa—the trade winds a voyager's highway between these two archipelagos at the heart of the Pacific. To leeward a smaller sailboat is headed in the same direction as us, pitching in the swell left from the gales of a few days ago. There is a tuna seiner farther offshore, pounding its way home in clouds of spray, and otherwise nothing—a bright morning with few clouds, the air unexpectedly dry after the concentrated wave of moisture that has just passed. I am admiring the impossible whiteness of our new mainsail against the blue sky while I walk around to inspect lashings and discuss plans with assorted members of the crew. With the wind on our beam the ship rides like it's on autopilot. A confident helmsperson can leave the wheel untouched for minutes at a time, the force of sails balanced in smooth equilibrium along the length of the vessel.

A sprinkling of nearly drowned mountaintops, Tonga sits at the northeast fork of an undersea range stretching 1500 miles back to New Zealand. Along this great seam the Pacific plate dives beneath the Australian plate, making a submerged trench seven miles deep with a boiling mass of magma in its lee, slowly burbling up to form new geography. At the northern end of this confluence slabs of crust are moving past each other at the ungodly rate of twenty-four centimeters a year—half again as fast as they do anywhere else on Earth. One night in 2019 a couple sailing their boat to Fiji were halted near here by a floating mass of pumice—buoyant gravel ejected from underwater volcanism—seven miles across. Their pictures show a small sailboat parked amid what appears to be a huge slab of concrete, with no water in sight.

The yachties were happily able to extricate themselves in due course, and scientists quick to offer that this was a not-at-all-uncommon phenomenon—perhaps a navigational hazard but also a natural mechanism for the spread of species across oceans. The rocks float from Tonga to Australia, gathering a slow aggregation of biology along their way. In this case no authority intervenes to be sure that these seaborne pebbles are cleaned of passengers before arrival, and crabs sprint happily ashore to their new home on the Great Barrier Reef. Miracles of nature aside, the thought of sailing headlong into a floating sidewalk is distressing. In my imagination I hear alarms beeping, the sound of footsteps on the ladder and someone pulling back the curtain to tell me our generators have shut down. With his compliments, the engineer would like me to know that our seawater plumbing is filled with tiny rocks. In another part of my brain I turn a page. I see no way to write a drill for this.

From Tonga we will sail gradually out of the tropics—west through the islands of Fiji and then south toward New Zealand. On a global weather map the latitudes we are bound for are broadly called the land of "westerlies," but as we've seen, the southwest Pacific is a good deal more complicated than that—patrolled as it is by the migratory hodgepodge of high- and low-pressure systems typical of the middle latitudes worldwide. In my casual way of tossing off comparisons I tell students to think of New Zealand as an austral California—one end sunny and subtropical, the other windy, damp, and never quite warm. This is true, if a bit oversimplified, since without the stabilizing influence of nearby continents there is rarely a true period of settled equilibrium. Auckland might be likened to San Diego, but any Californian forced to switch would be aggrieved by the frequency of wind and rain.

Another adjustment for northerners down under is that everything spins the opposite way in the southern hemisphere. The Coriolis effect torques wind to the left, not to the right, meaning that images you've had your whole life to paint are suddenly upended. Low-pressure systems (troughs, depressions, cyclones, and the like) rotate clockwise, while their high-pressure counterparts do the reverse. Cold fronts sprout from the bottom of the map rather than hanging down from the top. The change is scientifically logical but jarring for anyone trained in the boreal realm. Other fundamental frames of reference are similarly upended on long ocean passages—transits of the international date line, when today abruptly becomes tomorrow, or crossings of the equator, where the latitude scale on your chart suddenly starts counting in the opposite direction—but the reversing of the Coriolis effect manifests at a more visceral level. Imagine an American arriving at London's Heathrow Airport for the very first time and setting off down the left-hand side of the motorway in their rented car.

Columns of high pressure build in the atmosphere over the broad expanse of the Tasman Sea, drifting slowly east across New Zealand and onwards into Polynesia. In the *Mariners Met Pack*, a pithy Bible on weather in the southwest Pacific, Kiwi meteorologist Bob McDavitt calls these the "travelling anticyclones"—which to me sounds like the name of a band but in fact refers to the unmoored nature of these features relative to their staid counterparts in other oceans. The upper borders of the traveling anticyclones crowd the tropics, stirring the pot as their cool dry air displaces the warm and humid flow of the trade winds. Along such margins "squash zones" develop, abrupt pinchings of the pressure gradient that can increase the trade winds to gale strength with only subtle telltales on the weather map.

The general strategy of sailors in our position is to leave the tropics just after a traveling high has gone past, with the hope of riding the northerly winds in its wake. At least that is the idea. New Zealand is a thousand miles from Fiji, a passage long enough for anything to happen. Those on yacht voyages can seek the counsel of experts and wait for the perfect window. For ships like ours, obliged to stay on schedule barring some force majeure, there are still some helpful catechisms. Sail south, say those who know, and farther west than you'd think necessary—toward a point roughly two hundred miles north of New Zealand's uppermost extremity. Once the more predictable trades are lost astern, getting here at least preserves your options, with the assumption that whatever winds come next will be from somewhere in the west.

This is my first experience on this route, though I navigated half of it in reverse just a year ago, on a long cruise up from Auckland to a place called Raoul Island—a nature preserve where we weren't allowed to land but had an amusing fifteen-minute radio conversation with a park ranger named Oscar. He thanked us for the call and made sure we knew about the avocado trees and hot springs that dotted his private paradise. With a stiff wind on our beam we rounded Raoul's abrupt incendiary cone like a racing buoy and roared back the way we'd come, sailing from a brief pool of sunshine back into a murky overcast of teeming clouds. Throughout the night the wind built as a low-pressure system approached from the west, and we flew along like a rocket—8, 9, 10 knots, which is at least rocket territory for a heavy steel sailing ship. Underneath us the bottom was miles away, our sonar tracing the profile of an abyssal cliff as we traversed the lip of the Tonga Trench. It was an oddly disconcerting feeling, as though the profound depth would make a difference one way or another if things went wrong.

⌒

THE Kingdom of Tonga is the earliest known settlement in western Polynesia, where pottery fragments and cemetery DNA have dated village sites reaching back to 800 BC. This was the vanguard of a seafaring population that migrated south and east out of Asia, eventually to inhabit the entire South Pacific. On maps their intercrossed lines of expansion evoke a sort of centrifugal artwork, grown more intricate with each spin of the wheel. The Tongans speak English and are devoutly Christian but have been politically independent through their entire history— an enduring native monarchy in an ocean where every other island is a current or former territorial possession.

Vava'u, the northern island group of Tonga, looks like a jellyfish on the chart, a sprawled medusa raised into striking relief by geological uplift. This is in sharp contrast to Tonga's southern islands, flat as pancakes and parked on a completely different continental plate. The harbor here is perhaps the finest in the South Pacific, a deep landlocked basin surrounded by hills, invisible until our final turn in the channel—past a rusted buoy, through a curtain of rain and onto the dock, where in early morning the port is just coming to life. The customs inspector who comes to meet us is friendly, and generous with forms—after Christianity the most enduring mark of European influence on the world's far-flung societies. With cups of bad ship coffee we sit together at my tiny cabin table and flip through passports and declarations, the rain returning to drum on the skylight glass overhead. The inspector

stands to go, abruptly filling the space between me and the cabin
door.

"Thank you very much, Captain," he says. "Talitali fiefia—
welcome to Tonga."

Ashore in the village of Neiafu a single main street gives way
to neighborhoods of small houses, backed up against jungled hillsides with the distinct russet color of volcanic soil. Each dooryard
is a seeming race between plant species to see which can grow the
fastest: taro, breadfruit, stately mango trees, or the steady jungle
of bananas. Pigs are everywhere, sleek and exultant in their semi-
freedom. From a hilltop west of town it's possible to look south
down a chain of islands winding toward the horizon, their nubs
emerging from the ocean like pointed fingertips. Captain Bligh
was set adrift with his crew not far from here during the mutiny
on HMS *Bounty* in 1879. After a brief and lethal interaction with
natives on the nearby island of Tofua they did not stop again,
threading their way through the northern parts of Fiji and eventually to Timor in present-day Indonesia. In their open boat they
sailed 3500 miles, blown steadily west by the trade winds. Denied
charts by the mutineers, they made their own along the way, recording the first surveys of a blank piece of ocean now known as
the Bligh Waters.

Four years ago a new island emerged near Tofua in an abrupt
display of pyrotechnics, a steaming heap of cinders building an
isthmus between two smaller bits of existing land. Wind and sea
are now in the process of washing all this new terrain away,
though much more slowly than first expected. This has gotten the
attention of NASA geologists, who (I have learned) view marine
volcanoes as a useful proxy for features in more remote locations—
like Mars. NASA is an organization of space travel, but also one of
photography, ever in search of better ways to overfly strange terrain and glean as much information as possible in the process. Its

work in that way is not unlike oceanography, albeit with each exponent raised by multiple orders of magnitude.

"One key to doing science from space is trusting your instruments!" says Jim Garvin, chief scientist of NASA's Goddard Space Flight Center. Not an astronaut, he reminds me more than anything of a supercharged high school science teacher, enthralled with his subject and using exclamation points to end each sentence. The dry mountains of Mars, we are told, bear a more than passing resemblance to Pacific islands with the water taken away. This leads scientists to consider a possible connection. In observing the Red Planet, are we in fact looking at the stranded islands of a lost ocean? Is this where Earth might be headed, if rising levels of CO_2 tip some critical balance in the atmosphere? NASA is thus keen to study emergent island volcanoes, particularly the ones robust enough to endure for more than a few months. This subset is in fact rather small—including just a series of eruptions in 1958 at Capelinhos in the Azores, the newish island of Surtsey in Iceland, and now this spot, which the Tongans have called HTHH, Hunga Tonga–Hunga Ha'apai. By landing a team of scientists on a place like HTHH, Jim explains, a chance is offered to ground-truth the satellite sensors—thus permitting the critical assurance that what these exotic birds are seeing is in fact what's there.

Jim visited us at our shore facility during the preparations for this cruise. In front of our enthralled group he raced exuberantly through slides from the year previous, when the *Robert C. Seamans* made an initial exploratory stop at HTHH—her crew nervously briefed beforehand on all the potential hazards of new volcanic terrain. Nobody fell into a pool of boiling sulfur or disappeared in a sudden geyser of live steam, so we are going back for more. The first landing was in actuality a huge success, filling our office with newly drawn maps and triumphant photos of crews erecting instrument arrays on bare ashen landscapes. A satellite flying past made a priceless image of the ship at anchor, a fuzzy but distinct rice grain by the dark hulk of the volcano. Amid the records are also simpler visions, no less captivating, of life arriving in a brand-new place: Tendrils of green vine creep in photographs across black sand in a building net of foliage. A scurrying mouse, a perched owl. A coconut washed above the high-water mark, its shell half buried in soil and a sprout emerging upward into the light. Other people have been here, likely passing sailors and Tongans from nearby in the archipelago. It is possible in places to note their footprints and see that someone has collected a dozen of the sprouting coconuts and arranged them in a line. The web of human expansion extends in real time, setting out a grove of future palm trees that will be here to welcome visitors if this beach still exists twenty years from now.

At our berth in Neiafu I am having a late-day coffee with another NASA scientist named Dan Slayback. Dan looks *exactly* like an astronaut, slender and nondescript with a short salt-and-pepper haircut and the manner of someone too focused on details to give up much bandwidth for casual conversation. Jim Garvin's on-scene counterpart, Dan has flown here to meet us in Vava'u, his luggage overrun by plastic cases filled with precious instruments. We are discussing where to put all this gear as Penikolo's family arrives with the dinner they have prepared for the ship—a gastronomic catalog of everything it's possible to do with fish, taro, pork, crab, and coconut. There are perhaps sixty of us at mealtime, including many from Pen's extended clan who linger on the dock and must be urged aboard one by one. With the food they have brought we could easily feed a hundred, probably with leftovers for breakfast.

Dan was with the ship here a year ago and is excited at the prospect of doing more drone flights over HTHH—perhaps even at night to better capture the island's thermal signature with infrared imaging. He notes the importance of this measurement to volcanology. One must in this branch of science see the heat, not merely the light. How pitiful humans are, I think, to be limited by eyes that can observe but a thin slice of the electromagnetic spectrum. Captain Jay, in command for the previous visit, was a skilled drone pilot and thus a valuable ally in the overflight process. I confess to Dan that my own expertise lies in other areas, an expression that in the marine industry is a cherished euphemism for someone whose head is up their ass. What I need to tell Dan next is something more serious, which is that I've just gotten news of an emergency at home and will be leaving directly. The new captain, he will be glad to know, is in fact a superb drone pilot. With some shuffling of schedules I will rejoin the ship after Christmas, but in the interim our carefully laid plans have been quickly modified. Dan is sorry to hear this but responds with equanimity. It occurs to me that for someone in the spaceflight business a last-minute crew change is the mildest form of inconvenience. There are no exploding rockets here, after all—no canceled spacewalks or priceless probes impacting the Martian surface at ballistic velocity.

"Has this sort of thing happened before?" Dan asks.

I think this over and admit that yes, it happens fairly often, but I have myself generally been lucky. These are the Milanković cycles of a life at sea—logistics, weather, equipment, your own health and that of the others around you. Despite a long career my own personal emergencies have all to date transpired with the ship at the dock. Last before this was a day in 2012 when I awoke one morning in Honolulu with a bellyache and was rushed to the hospital for an appendectomy. Convalescing in a Waikiki hotel, I sat sipping noodle broth on my balcony and watched the ship sail off to sea without me.

I admit to Dan that I am nonetheless bitterly disappointed.

"Hopefully we'll all be back here again soon," he says.

What none of us realize at this moment is that we won't. In just over two years the tall conic mass of HTHH will begin smoking and then vanish in a single explosion, with a noise loud enough to be heard in Alaska. There will be a mushroom cloud the size of a hurricane—and when it clears, nothing left but a drowned ring of rocky fragments like the gap from a missing molar. On nearby islands whole villages are washed away by the ensuing tsunami, and a pall of ash covers the archipelago in a gray layer the color of an empty fireplace.

That is all it is possible to know at first. With Tonga's communications cable severed by the eruption, we will wait for weeks to confirm that our friends there are OK. Meanwhile the news will carry random signals of a remote place in mid-disaster: the woman who drowns while trying to rescue her dogs, the man who is swept out to sea and swims for a day to save himself. The pressure wave from the explosion is felt halfway around the world, a ripple in the atmosphere greater than any nuclear device has ever produced. Caught completely off guard, the scientists are gobsmacked. Imagine a tropical cyclone developing without warning in the backyard of the National Hurricane Center. Jim Garvin and his colleagues comment in the tone of meteorologists discussing severe storms, a hushed respect for tragedy cautiously applied over exultation at the pure power being revealed. There is no clean drinking water, and airplanes are unable to bring more, their runways eclipsed by toxic grit. Tonga is in this instant sent back in time, suddenly again a place that can only be reached by people in ships. From home I will look out in disbelief at the January snow and recall the photographs from our crew's first visit. The mouse and the owl, the new vegetation, the line of sprouting coconuts and footprints on a beach that will not be walked on again, at least not anytime soon.

Anyone wanting a long day of air travel should try the hop from Vava'u to Boston: First the long flight in a small plane to Fiji, a soporific three-hour drone across waters that until a day ago I thought I'd be traversing under sail. The tops of the clouds, not the bottoms, are washing past the oval windows with the sea far below. Next Fiji to Los Angeles, eleven hours. Six more hours to Boston, and then a bus, three hours more to my home in Maine. Through all this, thanks to the miracle of the international date line, I will arrive on the day I have left—the parameters of time, space, and speed distorted beyond recognition. One day, eight thousand miles. Perhaps fifty times what the ship might do in a fair strong breeze.

Between flights I send text messages back and forth to home, browsing for food in the terminals with my circadian clock utterly unmoored and no real sense of what meal I should be eating. The present crisis will happily resolve itself, though the world, we will all soon find, is standing into circumstances beyond the reach of anyone's expertise. As the vendors roll their shutters down for the night, I pace restlessly with a feeling that I know very well from the ship—the moments when, with bad weather coming, you realize that you have done all you can to get ready, your small world lashed down and the outcome yet unknown. All that's left is to wait.

13

YOUR CAREER OF CHOICE

2020

Cook Strait is the main dividing waterway of New Zealand, a flooded slot forty miles wide cut between the mountain ranges of the North and South Islands. From the middle on a clear day you can look to either side and see the hills split into steep relief by geological faults, above a stepped shoreline of old ocean bottom left high and dry by millennia of earthquakes. These are the Shaky Isles: an earthly Eden, temperate and progressive until a volcano erupts or an earthquake happens. Just as the geography of New Zealand can be compared to that of California, in the cheerful Kiwis I see more than a bit of the Californians themselves—delighted at their good fortune to live in a place where you can have your own avocado tree, all the while aware of the Faustian bargain they have struck in the process.

The only break in a tall mid-ocean land mass, Cook Strait is a laboratory for all that wind can do. Gales blow here just over 180 days a year, making it among the windiest regularly traveled bits of water in the world. Daunted by this prospect in my voyage planning, I sought out *The Cook Strait Mariner's Weather Handbook*, a slim work by a local meteorologist named Jim Hessel. Mr. Hessel has sorted through the many faces of Cook Strait and distilled them to a set of baseline scenarios for the nonscientist. Eager to learn, I dove in—only to find that even in simplified form the author saw thirty-eight potential weather situations for any given day in this complicated piece of ocean. At once awed and boggled, I set Mr. Hessel's remarkable work to one side and tried to conjure an even more basic mental model of my own.

The strait is a giant funnel, set to amplify whatever wind exists into something much more powerful—an effect comparable to what Captain Cook himself experienced in the Alenuihaha Channel on his fateful last voyage to Hawai'i, but on a greater scale and with more variability. The alternating realms of high and low pressure passing across New Zealand mean that there is nearly always a barometric difference between the east and west coasts— the wind racing through this lone gap in an attempt to balance the inequity. It was in Cook Strait that I first fully grasped the realities of a phenomenon called *lee troughing*—where even on an apparently fine day, the winds blowing past a high land mass will create an area of partial vacuum to leeward, building a declivity into which air charges from whatever direction it can. Mountains in wind, like rocks in a river, create untold turbulence in their shadows.

The longitudinal rifting of New Zealand's terrain channels air into rivers that then roar out at great speed anywhere that land meets water. Wellington, the otherwise delightful capital city, is at the bottom of one of these tectonic gullies. Here I once watched shipping containers blow off a dock and into the water like they were empty fruit cartons, bobbing around on the foaming surface while I tried not even to think about what the weather was like outside the harbor. Just east of Wellington is Palliser Bay, a broad inlet flanked by dramatic headlands. The place looks inviting until you realize there is nothing there—no wharf, village, or sign whatsoever of maritime activity. "Best avoided," say the sailing directions—confirming what the *Robert C. Seamans* herself found once several years ago when she wandered in and was promptly rebuffed with a storm-strength blast of wind from the north. Her only canvas, a tiny staysail built to withstand anything, was blown to smithereens.

Wellington was the site of a terrible shipwreck in 1968, when the ferry *Wahine* ran aground while entering the harbor in hurricane conditions. The size of an ocean liner, *Wahine* was caught by a 100-knot gust and spun sideways in mid-channel, grounding and capsizing just yards from shore. Fifty-two of her passengers perished amid efforts at rescue. This happened in April—early fall—through a set of circumstances that I found not too difficult to picture. A tropical cyclone had slid south down the slot between two traveling high-pressure systems, inhaling a deep gulp of antarctic air from a cold front pushing north along the same pathway. It was a storm of particular fury, though common enough in type to keep me on my toes as we make our own equinoctial approach to Wellington. These are the unbalanced forces that make the fall season so dangerous all around the world—the tropics, fat with heat and moisture after the long bright summer, and the poles, grown quickly cold amid the shortening of days. Yawning differences await resolution through the storms that follow.

Today the sky is an unrelieved blanket of gray clouds, making an easy job of it for anyone filling out the weather log. "ST," they will write, *stratus*, with an "⅞" following to indicate the fractional equivalent of sky that is covered.

"Why not use percentages instead of fractions?" the trainees sometimes ask.

"That would encourage unwanted individualism" is my answer.

The wind has come back up out of the south now after a few days of relative calm, carrying a damp chill that's driven everyone but the watch belowdecks in search of warmer alternatives. I am in my cabin, clicking through weather reports and eating licorice from the last of several bags purchased in Auckland. It is not yet clear whether my supply will last. I snack constantly at sea, browsing like a marmot on whatever is nearby as though the prodigious output from our galley is somehow not enough. I come home oddly thin, the calories perhaps burned away by stress, or more probably by ten thousand trips up and down the salon ladder for coffee.

On my computer screen assorted forecast files open and close, hands dealt out in a very slow card game where each is subtly different from the last. This is how it often feels to me, at any rate—every day a running comparison between what we have planned for the ship and what the weather has planned for us. On this voyage our best information arrives by text from the Meteorological Service of New Zealand and in something else called GRIB maps, received each morning over our narrow thread of email bandwidth. A key tool of at-sea weather routing, GRIB maps are made with gridded binary files—the raw output data of weather models—processed and shown on-screen as a field of colored arrows to indicate the strength and direction of predicted winds. Such is the bounty of the information world—the internal forces of the atmosphere converted to graphics and displayed in all their pixilated glory. With a few minutes of scrolling you might find something similar on your phone, a miraculous direct channel to a supercomputer roaring away in some guarded basement. There is amid this abundance a need for caution, as the data in these maps come without the touch or interpretation of any meteorologist. No Joe Sienkiewicz or Frank Musonda has yet waved their hand in blessing over these weedy graphics. They are but a single stream of information, just one of the many used to build actual forecasts after subjective consideration. Caveat emptor.

Someone is in the doorway.

"Is your licorice going to last until Wellington?"

It is the steward, my most frequent visitor for nonnavigational matters.

"We'll see."

It is time to talk again about purchasing provisions, each port call representing as it does the chance to fill our ever-draining larder of stored calories. With the upbeat composure of someone at a press briefing or real estate closing, Sabrina finds a secure place on the settee and makes her knees into a desk for her clipboard. This is our fourth trip together, or perhaps the fifth. Under her helmsmanship the galley accomplishes the incomprehensible day upon day, conjuring meals for forty from a heaving stove and freezers jumbled into chaos. Bread emerges from the ovens as though delivered from Paris. I hear raised voices on deck above us and then the percussive thud of an out-of-step wave. There is a gurgling sound and a rippled sheet of water across the cabin skylight. A dribble finds its way in somewhere and spatters across the floor.

Sabrina runs through the last items on her list and stands to go, grasping a bookshelf for balance.

"Would you mind if I keep dinner in the galley, for people to come and help themselves?" she asks. "I'm happy to set out the serving trays, but I doubt that they'll stay on the tables long enough to eat from."

"That sounds good to me," I tell her. "Nobody knows better than you."

Weather can catch you by surprise, as with the proverbial wall of wind that turns a calm day into a sudden gale. It can also escalate incrementally, each hour marginally worse than the last until suddenly you are making your way through breaking waves six meters tall, walking out of the chartroom to realize that all you can see is water unless you look straight up. This is now by any measure another gale, our second in a week, but not of the sort that has earned much mention in the forecasts. The local weather at the moment in fact looks lovely, on paper. A dome of high pressure has settled firmly over New Zealand, creating fine late-summer conditions ashore. No doubt in Auckland the locals are watering their gardens and loading cars for a weekend at the beach, but here, two hundred miles away at the rim of the anticyclone, we are parked on the edge of an atmospheric cliff—taking a cold wind in the teeth as the diverging air spins away and back toward Tonga.

An instructor named Mark Schwarz at the Meteorological Service of New Zealand has given me an insightful take on why these days that look so nice on the map are not necessarily tranquil for all: The dense stable air of a high-pressure system can squeeze the air below it like a lid, sending it whirling off at the edges and out through gaps in the coastal terrain at double speed. The sinking air acts like the roof of a tunnel, while any land in the way forms a wall. Some of Mark's consulting work involves weather support for rocket launches, on the North Island's east coast at a place called Mahia. Wind is a big issue here, particularly wind *shear*, the sudden change in velocity between different air layers. Significant wind shear means a no-go for launch. Picture running at supersonic speed with a broom handle balanced on your fingertips, says Mark. That is rocket science.

It is a long, rough ride for us now to Wellington, an objective we strive toward in a steady slog, adjusting our course so that we wallow up and over the biggest waves with only the occasional heart-stopping thud as we meet one head-on. We progress slowly, at walking speed. Slower. Wedged into my bunk, I have over the last few days been reading a book about the Battle of the Atlantic. Merchant ships from North America are fighting to reach England through endless weeks of storms, hunted by packs of U-boats. All this, I think, plus torpedoes.

The Cook Strait Mariner's Weather Handbook advises against approaching the strait in southerly gales, when the topography and outsized tidal currents combine to create some of the worst possible conditions. This said, the short interval afterwards may provide the best chance of all, a window in the brief calm that follows each storm. In past years I have only imagined this, but as we make our way inshore a day later, the sea transforms itself again into an uncommon smoothness. The mountains of the South Island, reminiscent of the Sierras, are backlit in crisp outline at sunset. Wellington, still fifty miles away, paints its faint glow on the opposite horizon. This is perhaps my favorite port city of all, a hilly place of coffee shops and green parks where the locals stride gamely about, leaning steadily into a wind that never seems to stop. Tonight is thus a startling anomaly—calm enough to fish with our plankton nets and watch dolphins towing trails of bioluminescence past our bow wave.

The chief scientist, on perhaps his fortieth cruise and no stranger to suffering, stands next to me at the rail as the mountains fade finally into dusk. A laboratory team is organizing equipment for more plankton-catching, their motions still wary as though the seas might suddenly return.

"That," says the chief scientist, "is I think the roughest week at sea I have ever had."

"Your career of choice," I offer.

"Hm," he says. It counts as a full sentence for someone from Finland.

Jan has moved away from Scandinavia for a career as a biological oceanographer. We are exactly the same age but finished college in different years as a consequence of his compulsory military service, a time that included learning to ski and shoot amid live fire exercises in wintertime Lapland. The Finns are serious about staying ready to fight the Russians. His subsequent life abroad has been motivated, in his own words, by curiosity, and perhaps a long-satisfied appetite for cold places. Behind us the second mate is digging through the lazarette, a small hold behind the steering station where our mooring gear is stowed. This typically is a job for morning, but after the shaking we have endured, some doubt exists as to the order of our equipment—the bulky coils of line, heavy rubber fenders, and crates full of hardware likely tossed now into a tangled morass.

"I'm going to sort this problem before it's a problem," she has just told us, finishing her tea and descending into the hatchway. Here is seamanship, I think, for any who would want an example—the compulsive instinct to thwart mishaps in advance, thus depriving them of the chance to cascade into some greater unknown, large or small.

<hr />

THE Meteorological Service of New Zealand resides in a boxy glass building at the edge of a botanical garden, overlooking the broad bowl of Wellington harbor. Save for a generous assortment of roof antennas, it might be mistaken for a bank headquarters. Several years ago, with the ship held up by a spell of particularly bad weather, our crew knocked on their door and asked to be let in for a tour.

"Oh, no," we were told, "we don't give tours, mate. We're just a bunch of nerdy people looking at computer screens anyway. You wouldn't be interested."

Well, that, we offered, is where they were wrong.

The staff at the Meteorological Service are a cheerful crew, poached from the ranks of aeronautical engineers and particle physicists at local universities. Theirs is a demographic unthreatened by mathematics. It couldn't be any other way, as they have their work cut out for them in making sense of the atmospheric circus that they live within—the humid tropical air pushing down from the north, clashing with great chilly blasts off the Southern Ocean and occasional injections of real cold from Antarctica. To the west, the vast dry expanse of Australia slings its superheated air across the Tasman Sea, where it gets good and wet before crashing into the tall mountains of the South Island. Almost anything can happen, on any day of the year. If you are sailing in New Zealand, you live by the three-day coastal forecast, which is renewed every twelve hours and never the same twice in a row. The service has some confidence in their seventy-two-hour prediction, but as one of their people once told me in an aside, speculating any further ahead than that would be time better spent watching rugby at the pub.

We have been welcome visitors at this friendly agency ever since our initial gate-crashing three years ago—though in an ominous sign of how things are about to change there is today a bottle of hand sanitizer by the reception desk, and a message from Mark Schwarz to say that we'll be meeting in a room downstairs instead of up on the main forecasting floor. It is March 16, 2020. New Zealand, ever optimistic, has nonetheless begun to look over its shoulder at events overtaking the rest of the world. Mark appears moments later to greet us, along with a colleague from the training staff named David Webster. Together they look like a heavy metal guitarist and his accountant, both still a bit unbelieving that we'd all take two hours from our day and walk up a steep hill to pay them a visit.

David and Mark agree that this is a great place to be a

meteorologist. The southern hemisphere—short on land and thus lacking weather stations—was the poor cousin of this science until the advent of scatterometry made it possible to measure wind from satellites with radar scans of the sea surface. Suddenly great swaths of ocean were there to behold, covered in clean unpolluted air and just waiting to be studied. Today Mark talks with us about the transport of water in the atmosphere, emphasizing how clouds can't form effectively without a supply of airborne particles to act as nuclei for condensation. In fact, he asserts, perfectly clean air can require as much as 800 percent humidity to yield water droplets spontaneously. Southern Ocean storms rely on a steady supply of airborne sea salt for their condensation nuclei, which turns out to be very efficient for this purpose. Land-based particles—bits of clay, dust, and man-made pollution—are involved also, as are tiny bacterial organisms. Perched upright on his stool, Mark explains how bacteria have been found to migrate between habitats by a remarkable mechanism: Riding the dust of decayed matter, they are blown aloft by convection and deposited downwind in raindrops, onto new surfaces where the arriving microbes can then flourish on fresh media. Sailing on their cloud ships from field to meadow, bacteria are the primordial navigators of our biosphere.

Outside the windows a tall row of cumulus pushes past across the sky, the harbinger of another advancing front that will drive temperatures down into sweater territory and the winds back up to storm strength by evening. As we speak, the angled topography of the South Island is forcing this dense tongue of cold air up the coast, soon to reach Cook Strait in a freshly charged subpolar blast. Our calm sunny day, only hours old, ends abruptly. In the harbor below us a vessel turns in the fairway and heads back out to sea. Today New Zealand has closed itself to cruise ships and required that all arriving airline passengers quarantine in the hope of limiting the spread of the COVID-19 virus. This will not work, and in three more days the whole country will seal its border to noncitizens. Wellington strives to remain its breezy upbeat self as the steady drip of alarming news from abroad amplifies, but telephones are ringing. On my way back to the ship I meet someone on the sidewalk carrying a double armload of toilet paper. Uncharacteristically, they do not say hello.

Docked just north of the *Robert C. Seamans* is a small French cruise ship that has just canceled the remainder of her current voyage. Her passengers, abruptly evicted, are flowing out onto the wharf with the stunned belligerence of interrupted vacationers. There are taxis milling about the perimeter, valets wheeling baggage carts, and staff handling paperwork at a table by the gangway. It all looks like the lobby of a busy hotel with the building lifted away. On our own ship the engineer is busy talking to a fuel company on his mobile phone. A delivery had been promised for tomorrow, but suddenly the plan is no longer firm. From our home office comes a cascade of new messages. Plans have tipped suddenly into flux on the sudden arrival of this new and different form of weather event, and logistics are being rushed into place to end our own cruise early and prepare the ship for an expedited return to the US. Our final port call, scheduled for Christchurch, is canceled.

When I gather the group to share this news, there is not dismay but a rapid switch into problem-solving, the click of engaging gears nearly audible.

"Got it," says someone. "Time to run the global pandemic drill."

"Yes, exactly," I tell them. "You'll find it right on the station bill with all the others."

Pens scratch across clipboards, and submeetings sprout spontaneously in various corners. Five minutes later I see that two of the scientists are moving equipment out of the winch-house to make room for the now-imminent tide of off-going personal gear. Someone is putting out extra mooring lines in preparation for the rising wind, and the engineer comes to find me where I've stopped to stare off into space.

"Do you have a minute, Cap?"

A strapping and polite former coast guardsman, Henry is a dimensional opposite to Danger Dave, with whom I first sailed off into the Pacific years ago. There is no piece of gear aboard that he cannot rebuild into an improved condition.

"My day is made of minutes," I tell him.

"Our fuel is on its way. We're finished with the main engine service and ready to move the ship at any time if needed."

"Excellent. I haven't heard anything from the harbormaster, so I'm hoping we are OK at this dock for now."

"Oh, and snack is ready in the salon."

"Even better."

Henry salutes with a hint of frivolity and turns to go.

Once, at an end-of-cruise party, I was given a farcical award, on a certificate made from a paper plate: "Best person," it said, "to remove your appendix at sea with kitchen utensils."

I was flattered, but if I needed this done myself, I would without doubt go looking for Henry or someone like him, members of an adaptive species native to the world of ships. If I am ever in the jaws of some utterly intractable and unseen problem, I think, send me a sailing crew and they will figure it out, with time left to remember that they are hungry.

My relief, traveling sooner than first planned, has by his own resourcefulness found a seat on what will turn out to be the last US flight permitted into New Zealand. When I check the status of my own flight home, it is overwritten with a concerning message on the airline's web page. "Call us," it advises, as though one even could. Elsewhere on the internet flights are disappearing as

I watch or updating with their prices startlingly inflated. In a wild stroke of luck, I'm suddenly able to change airlines and buy a seat on the last Air New Zealand flight back to America. I've never seen anything quite like this. I have for some time been intrigued by the idea of moving to Wellington, but this is not how I'd envisioned it.

The new captain arrives the next morning and we go hastily through papers. Many of the other arriving crew are thankfully already in the country and they trickle aboard while we are talking, dragging the detritus of their own truncated holidays. As I climb into my taxi, there are still people out walking, but the characteristic buoyancy of Wellington is gone, the parks deserted and high tables standing in empty rows under café awnings. It is a strikingly clear day, with a stiff breeze scouring the harbor surface into corduroy—perhaps the loveliest weather since we arrived. The crew will quarantine here aboard the vessel for two weeks before departing, kept busy by the endless list of chores that every ship carries. Allowed ashore only for trips to the grocery store and solitary exercise, for amusement they fill the sidewalk on the wharf with colorful graffiti done in chalk. On their last day someone draws a big heart inscribed with the words: "Thank you, Wellington!! We'll be back soon!" The next morning they are gone, started on the six-thousand-mile voyage that will take them home to Hawai'i.

14

IN THE MONTH OF FARCH

Karen is waiting for me at the bus station. As the coach rolls through the parking lot, I spot her car, see the peak of her jacket hood and the distinct profile of our feral Scottish terrier, suddenly alert behind the window glass. There is still snow on the ground, and from the terminal comes a great blast of stale heated air as we carry our bags away from the bus and through the sliding doors. This is what I notice first when coming home from milder latitudes—not cold but heat, the sense of air trapped in closed spaces and made warm with boilers. From the city we drive east, the snow cover growing more consistent and the sky opening. Northern New England is still far from real spring. The weeks of alternating overcast and sunlight, the sharp dry cold and sneaky damp, will join into their own attenuated semi-season, our coat hooks crowded with every possible variant of outerwear. It is pointless, a friend once told me, to give the months different names at this time of year. We should instead adopt a new combined term, one more sincere and less giving of false hope. February and March would become simply *Farch*. I could only agree.

My enormous bag is jammed behind the seats of our car, blocking out the sun. Up a back road rutted by frost is our little house, its tall front windows pointed hopefully toward the south. Inside, the dusty light falls on shelves of books and memorabilia—ivory carvings from Greenland, a woodcut purchased long ago in the Seychelles Islands. A sooty piston, taken from a ship's engine for replacement after untold millions of revolutions. In a dish of keys and pocket change on the counter are odd coins from Cuba, Kiribati, and the Azores. I built the first rooms of this place in my mid-twenties, imagining a modest base camp for an itinerant sailor and his few possessions. Quickly afterwards there were two of us, inhabiting a space designed for one but living happily nonetheless—perhaps conditioned by our shared experience in much smaller quarters.

I dump my gear out in a pile and put water on for tea. The dog pounces on some stray item and disappears under a chair. In the weeks to come I will stand watch here each morning at my kitchen table, in communication with two other captains who are assisting the ship as she makes her departure from New Zealand and hastens home—first eastward across the Roaring Forties, then north through the trade winds to Hawai'i. A six-thousand-mile voyage, taken nonstop to skirt the unfolding complexities of landing abroad. Two Dutch sailing ships, the *Europa* and *Wylde Swan*, are embarked on parallel odysseys in the Atlantic. Caught far away by the unexpected, they also have elected to bring their lines aboard and simply sail for the Netherlands, embarking on the unplanned adventures of a lifetime.

For two months these three ships ride alone in their worlds, their crews momentarily safer than anyone else on the planet from the pandemic. On my laptop I watch the daily animations of weather models, sharing messages with someone at our office who is in touch with the *Robert C. Seamans*. On my screen the vast Pacific is eleven inches across, the swirling graphics of wind like rapids on a river. In watching, I feel the pull of currents as I would on my own tiny kayak, understanding the importance of staying where the flow is in your favor and knowing the danger of being caught unaware by sudden reversals.

In paragraph form the ship's mission is this: Sail east, keeping north of the antarctic low-pressure systems to assure fair winds, while looking carefully back at the tropics for signs of late cyclones. Somewhere near the Austral Islands—the remote southern frontier of French Polynesia—turn left and head for Hawai'i.

From this pivotal corner, the track wanders across the shifty subtropics and back into the trade winds. Here like a dragster under full sail the *Robert C. Seamans* rockets north, the wind warm on her quarter and the low atolls of the Tuamotus passing invisibly by to either side. The miles fly by, sent back in exultant daily totals each day by the crew: 150, 175 . . . 199! They are on the ride of their careers. Ashore in our kitchens and offices, the rest of us are like astronauts in mission control—thrilled to hear this news, sorry the adventure is not ours. Grateful to be home while the world sails on, into a storm for which the models are not yet written.

CHAPTER NOTES

1. THE HOURLY

For historical background on the metrics of weather applied by observers at sea, I looked first to Scott Huler's *Defining the Wind: The Beaufort Scale, and How a 19th-Century Admiral Turned Science into Poetry* (New York: Three Rivers Press, 2004). Verification of basic scientific parameters (in this and subsequent chapters) came primarily from Donald Ahrens and Robert Henson's *Meteorology Today: An Introduction to Weather, Climate, and the Environment*, 12th edition (Boston: Cengage, 2019) and *Reeds Maritime Meteorology*, 4th edition, by Maurice Cornish and Elaine Ives (London: Bloomsbury, 2019). The details of weather's moving parts—waves, fronts, and mesoscale features like the sting jet that overtook Captain Jay Amster on the *Robert C. Seamans* in 2019—were largely resolved in the twentieth century and are often best explored in journal accounts from the researchers themselves (see Selected Articles). The image of grain ships thundering east across the Southern Ocean is a universal one in maritime mythology, but for a tale of actual events I have found few more compelling than Basil Lubbock's *The Last of the Windjammers*, volume 1 (Glasgow: Brown, Son, and Ferguson, 1927), written in a time when square rig sailors were still a regular part of the working waterfront populace.

2. FIRST PRINCIPLES

Technical discussions in this chapter are supported by information from the texts named in Chapter One, along with material from W. L. Ferrel's *The Motions of Fluids and Solids Relative to the Earth's Surface* (New York: Ivison, Phinney, 1860)—a mixed volume of dense mathematics and predicted atmospheric behavior, all miraculously derived before the election of Abraham Lincoln.

It is hard to set sail in the Pacific without considering the trails left—for better or worse—by Captain James Cook and his many followers. As remarkable as Cook's navigational feats were, they may have been equaled by his skills as a diarist, and anyone wanting a real-time narrative need only procure a copy of his recounting, *The Journals of Captain Cook*. On my bookshelf is the 1999 Penguin paperback edition, edited by Philip Edwards. My own perspectives as a foreign mariner in today's Pacific were also influenced by several readings of Tony Horwitz's *Blue Latitudes: Boldly Going Where Captain Cook Has Gone Before* (New York: Picador, 2002)—a nuanced and often humorous journey into the legacies of colonialism, for both colonists and the colonized.

3. THE CLOUD FOREST

In tandem with a review of clouds and their meteorology, my discussions of sudden wind events and their threat to sailing ships are much informed by the work of Captain Daniel Parrott in *Tall Ships Down: The Last Voyages of the* Pamir, Albatross, Marques, Pride of Baltimore, *and* Maria Asumpta (Camden, ME: International Marine, 2003). There are few prospects more frightening to a sailor than the thought of having your ship blown down and sunk, and I am grateful to those among my colleagues who have been willing to share their own difficult stories with the hope of fostering safer outcomes in the future. (See Selected Articles.)

4. THE SERPENT'S COIL

This chapter owes much credit to Peter Moore's *The Weather Experiment: The Pioneers Who Sought to See the Future* (London: Vintage, 2015), for information on the early conceptualization of cyclones among nineteenth-century mariners and scientists. The remarkable emergence of quantitative meteorology under the Norwegians is described in Robert Marc Friedman's *Appropriating the Weather: Vilhelm Bjerknes and the Construction of a Modern Meteorology* (New York: Cornell University Press, 1989), and uniquely summarized in Jacob Bjerknes's original monograph on wave cyclones from 1926 (see Selected Articles). For discussions of maneuvering a ship under threat from a tropical cyclone—and Captain Phil Sacks's successful evasion of Hurricane Frances aboard the schooner *Westward* in 1992—I turned to *Auxiliary Sail Vessel Operations for the Professional Sailor*, 2nd edition, by Captain G. Andy Chase (Centreville, MD: Cornell Maritime Press, 2016). The inherent risks of crossing a hurricane track at sea are revisited by Captain Chase in his article "Lessons of the *BOUNTY*" (see Selected Articles).

5. GOD'S ROOF

My own education in the remarkable accomplishments of Pacific navigators is owed in no small part to Christina Thompson for her recent book, *Sea People: The Puzzle of Polynesia* (London: William Collins, 2019). Also valuable were Andrew Crowe's *Pathway of the Birds: The Voyaging Achievements of Māori and Their Polynesian Ancestors* (Honolulu: University of Hawai'i Press, 2018) and Sam Low's *Hawaiki Rising: Hōkūle'a, Nainoa Thompson, and the Hawaiian Renaissance* (Honolulu: Island Heritage Publishing, 2013). One of America's present experts on wayfinding is in fact a particle physicist by trade, and I thank Dr. John Huth of Harvard University for both his book—*The Lost Art of Finding Our Way* (Cambridge, MA: Belknap Press of the Harvard University Press, 2013)—and his enthusiastic collaboration during my time with the SEA Education Association of Woods Hole. It was also during my SEA voyages that I had the good fortune to sail briefly with Captain Ka'iulani Murphy and some of her fellow navigators at Hawai'i's Polynesian Voyaging Society, modern stewards of an ancient art.

6. THE MOTHER SHIP

For stories of early marine meteorology and its founding practitioners—Beaufort, FitzRoy, Redfield, and the like—I again credit Peter Moore's *The*

Weather Experiment. Discussions of modern forecasting were informed by Andrew Blum's *The Weather Machine* (New York: Ecco, 2019) and a monograph by longtime National Weather Service scientist Dr. Harry R. (Bob) Glahn: *The United States Weather Service: The First 100 Years* (Rockville, MD: Pilot Imaging, 2012). A gripping image of the service's formative years comes from Erik Larson's *Isaac's Storm: A Man, a Time, and the Deadliest Hurricane in History* (New York: Random House, 2000), while insights on forecasting during World War II were provided by John Ross in *The Forecast for D-Day and the Weatherman Behind Ike's Greatest Gamble* (Guilford, CT: Lyons Press, 2014). In supplement to these books was a long list of articles from the *Bulletin of the American Meteorological Society* (*BAMS*), a trove of information for any researcher on this topic (see Selected Articles). Most of all, thanks are owed to Joe Sienkiewicz, a longtime friend and resource to all of us who travel under sail—slowly, sensitive to weather, and with precious cargo.

7. VOLTA DO MAR

Perspective on my own Mediterranean sailings was expanded by readings of Alfred Crosby's *Ecological Imperialism: The Biological Expansion of Europe, 900–1900* (New York: Cambridge University Press, 1986, 1993, 2004). Also, Nigel Cliff's *The Last Crusade: The Epic Voyages of Vasco da Gama* (New York: Harper, 2011), Laurence Bergreen's *Over the Edge of the World: Magellan's Terrifying Circumnavigation of the Globe* (New York: HarperCollins, 2004), and Robert Fuson's *Juan Ponce de León and the Spanish Discovery of Puerto Rico and Florida* (Blacksburg, VA: McDonald and Woodward, 2000). In present-day Spain I owe Nick Lloyd for offering me a greater grasp of Spanish history than a passing mariner might ever hope to have, and for his book, *Forgotten Places: Barcelona and the Spanish Civil War* (Nick Lloyd, 2015).

8. SAFELY OUT TO SEA

My discussions of the human element in marine casualties were supported by another book from Captain Dan Parrott, *Bridge Resource Management for Small Ships* (Camden, ME: International Marine, 2011). This illuminating set of case studies is informed by Captain Parrott's many years of experience at sea and as a professor at Maine Maritime Academy. Also insightful was Malcolm Gladwell's "The Ethnic Theory of Plane Crashes," in *Outliers: The Story of Success* (New York: Little, Brown and Company, 2008). Official investigation reports for the *Picton Castle, Bounty,* and *El Faro* casualties are referenced in Selected Articles. Reading the stories of others who have gone to sea and met with misfortune is a humbling process. One can only accept this information in the way its authors intended—which is with the hope that the lessons learned may improve the safety of future voyages.

9. A RIVER OF WIND

A full account of the mysterious disappearance (and rediscovery) of the airliner *Star Dust* is to be found in *Star Dust Falling: The Story of the Plane That Vanished,* by Jay Rayner (New York: Doubleday, 2002). For the emergence of the jet stream as a core principle in atmospheric science—and its modern role at the heart of forecasting and climate change—I again mined the abundant vein of scientific articles that exist on the topic, beginning with Carl-Gustaf Rossby's original work at the University of Chicago. (See Selected Articles.) Special thanks to Dr. Jennifer Francis of Woods Hole's Woodwell Climate Research Center, for her willingness to review early drafts of this chapter and improve my understanding of certain key points.

10. CLAMBAKES OF ANTIQUITY

For background on the underpinnings of El Niño, I referred to Edward S. Sarachik and Mark A. Cane, *The El Niño–Southern Oscillation Phenomenon* (New York: Cambridge University Press, 2010), as well as Sir Gilbert Walker's original monograph, *Correlation in Seasonal Variations of Weather, IX—A Further Study of World Weather* (Calcutta: Government of India Press, 1924). Again, credit is owed to Andrew Crowe's *Pathway of the Birds* for his thoughts on how interannual wind variations might have shaped the journeys of island voyagers. Thanks also to my many friends in the world of oceanography who built my understanding of ENSO in uncounted lunchtime lectures and long nights on deck spent watching gear deployments.

Special appreciation to Professor Daniel Sandweiss at the University of Maine, for taking time to discuss the extensive work he has done in South America with others from the Climate Change Institute.

11. THE ICE DRILLS

On the development of pilot charts, naval hydrography, and the long inter-play between climatology and navigation, I looked to Chester Hearn's *Tracks in the Sea: Matthew Fontaine Maury and the Mapping of the Oceans* (Camden, ME: International Marine, 2002), as well as Nathaniel Philbrick's *Sea of Glory: America's Voyage of Discovery—The US Exploring Expedition* (New York: Penguin, 2003). For an overall history of polar exploration, I tapped an old favorite, Pierre Berton's *The Arctic Grail: The Quest for the Northwest Passage and the North Pole, 1818–1909* (New York: Viking, 1988). Also valu-able was Ken McGoogan's *Fatal Passage: The Story of John Rae, the Arctic Hero Time Forgot* (New York: Carroll & Graf, 2001). For an account of the more recent (if also nearly forgotten) adventure of the SS *Manhattan*, I read Ross Coen's *Breaking Ice for Arctic Oil: The Epic Voyage of the SS* Manhattan *through the Northwest Passage* (Fairbanks: University of Alaska Press, 2012). Back-ground on ice cores and their foundational role in paleoclimatology came from Paul Mayewski and Frank White's *The Ice Chronicles: The Quest to Un-derstand Global Climate Change* (Hanover, NH: University Press of New England, 2002). Dr. Mayewski—who is director of the Climate Change In-stitute at the University of Maine—was of great help in modernizing my understanding of the field after my long-ago meeting with his GISP-2 re-search team in Greenland. Other important sources for this chapter in-cluded Brian Fagan's *The Little Ice Age: How Climate Made History* (New York: Basic Books, 2000, 2019), and Bob Drury and Tom Clavin's *Halsey's Typhoon: The True Story of a Fighting Admiral, an Epic Storm, and an Untold Rescue* (New York: Grove, 2007)—each a landmark account of the recurrent crossings between weather and history. The quote from Percy Langston re-garding sea level rise on Kiritimati is taken from the video documentary *Between Sky and Ocean* (https://youtu.be/aXrZ-k333aw), made by Wojciech Hupert in 2012, not long after my own visits to the island in the *Robert C. Seamans*. For a frightening and informative personal memoir of the nuclear tests at Kiritimati, I recommend J. Haggas's *Christmas Island: The Wrong Place at the Wrong Time* (London: Minerva, 1997).

As a baseline resource for this chapter, I used Robert Henson's *The Thinking Person's Guide to Climate Change* (Boston: American Meteorological Society, 2014). Books can't be written fast enough to track the changes currently at play in this field—particularly in the polar regions—and here I relied extensively on journal articles to keep pace with events. (See Selected Articles.)

The ultimate form of this chapter was shaped greatly by conversations with my old friend and shipmate Dr. Kevin Wood—scientist, craftsman, polar voyager, author, husband, and father. Sadly, Kevin passed away in February of 2022, just after this manuscript was finished. No doubt his substantial contributions to science will be carried forward by colleagues, but for many Kevin's departure will leave a gap that cannot be filled. After some deliberation I have decided to leave him in these pages as he will be remembered: alive, in the present tense, filled with enthusiasm for the world's interesting things and always ready to share new ideas.

12. HTHH

Much of what is told here of Tonga and Samoa is taken from conversations I had while going through my days in these two distinct chambers of Polynesia's heartland. Once, after a flight from Honolulu to Pago Pago (there are two per week), I was approached in baggage claim by the man who'd been seated next to me in coach. Did I have a ride? A place to stay? If not, I was welcome to spend the night with his family—hospitality offered at a level that was touching but not at all uncommon. The artists Regina and Su'a Fitiao have offered such a welcome to the entire SEA crew across an arc of multiple visits—turning American Samoa into a second home for many, just about as far as it's possible to go from the mainland.

In approaching the dog's breakfast that is weather in the southwest Pacific, one might begin with Bob McDavitt's *Mariners Met Pack: South West Pacific* (Auckland: Captain Teach Press, 2007)—a fine book, which, rather like a road map of Boston, constitutes a mere introduction. To glean any real understanding, it is necessary to drive into the place and get lost—engaging in serial acts of self-rescue until some actual cognitive model begins to emerge.

13. YOUR CAREER OF CHOICE

What literacy I possess in Kiwi weather owes much to the slim guides of Jim Hessel (Auckland: MetGen Meteorological Consulting, 1991) and the gracious welcome given to our ship's staff on repeated visits at the Meteorological Service of New Zealand, in Wellington. Special thanks to Mark Schwarz, who always found the time to greet us on what must have been a long string of busy days. Thanks also to David Webster, Neal Osborne, Ross Bannister (who made sure our barometer was telling the truth), and Ian Attwood—the man who first admitted our band of hopeful gate-crashers on a morning too foul for sailing.

14. IN THE MONTH OF FARCH

No research was required here to know that there is no better part of voyaging than coming home.

SELECTED ARTICLES

Allen, D. R., 2001. "The genesis of meteorology at the university of Chicago." *BAMS*, vol. 82, no. 9.

Askenov et al., 2017. "On the future navigability of Arctic sea routes: High-resolution projections of the Arctic Ocean and sea ice." *Marine Policy*, vol. 75, 300–317.

Barnston, A., 2014. "How ENSO leads to a cascade of global impacts." *ENSO Blog* (Climate.gov), https://www.climate.gov/news-features /blogs/enso/how-enso-leads-cascade-global-impacts.

Bates, C., 1956. "Marine meteorology at the U. S. Navy hydrographic office—a resume of the past 125 years and the outlook for the future." *BAMS*, vol. 37, no. 10.

Bjerknes, J., 1966. "A possible response of the atmospheric Hadley circulation to equatorial anomalies of ocean temperature." *Tellus* XVIII, 4.

Bjerknes, J., and H. Solberg, 1922. "Life cycle of cyclones and the polar front theory of atmospheric circulation." Bergen Geophysical Institute.

Byers, H., 1960. "Carl-Gustaf Arvid Rossby: A Biographical Memoir." National Academy of Sciences, Washington, DC.

Carlowicz, M., 2003. "A river runs through it: Chronicling the currents of the North Atlantic." *Woods Hole Currents*, vol. 10, no. 2.

Chase, G. A., 2013. "Lessons of the BOUNTY." *WoodenBoat*, no. 233.

Francis, J., 2019. "Yes, climate change is making severe weather worse." *Scientific American*, https://www.scientificamerican.com/article /yes-climate-change-is-making-severe-weather-worse/.

Francis, J., and S. Vavrus, 2013. "Evidence linking Arctic amplification to extreme weather in mid-latitudes." *Geophysical Research Letters*, vol. 39, L06801.

Headland et al., 2020. "Transits of the Northwest Passage to end of the 2020 navigation season." Scott Polar Research Institute at Cambridge, http://thenorthwestpassage.info/scott-polar-research-institute.

Hugonnet et al., 2021. "Accelerated global glacier mass loss in the early twenty-first century." *Nature*, vol. 592, April 29, 2021.

IRI, 2015. "2015 El Niño: Notes for the East African malaria community." International Research Institute for Climate and Society, New York.

Koračin et al., 2014. "Marine fog: A review." *Atmospheric Research*, vol. 143, 142–175.

Law et al., 2010. "Plastic Accumulation in the North Atlantic Subtropical Gyre." *Science*, vol. 329, no. 5996, 1185–1188.

Lewis, J., 2003. "Ooishi's observation: viewed in the context of jet stream discovery." *BAMS*, vol. 84, no. 3.

Little, T., 2019. "In Greenland village, shorter winters cast doubts over dog sledding." Phys.org, https://phys.org/news/2019-09-greenland-village -shorter-winters-dog.html.

Lorenz, E., 1969. "The Predictability of a flow which possesses many scales of motion." *Tellus* XXI, no. 3.

Marshall, A., 2019. "Icebound: The climate-change secrets of 19th century ship's logs." Reuters, https://www.reuters.com/investigates/special -report/climate-change-ice-shiplogs/.

Mayewski et al., 2014. "Holocene warming marked by abrupt onset of longer summers and reduced storm frequency around Greenland." *Journal of Quaternary Science*, vol. 29, no. 1, 99–104.

Middleton, W. E. Knowles, 1944. "A brief history of the barometer." *Journal of the Royal Astronomical Society of Canada*, vol. XXXVIII, no. 2, February 1944.

Mikesh, R., 1973. "Japan's World War II balloon bomb attacks on North America." *Smithsonian Annals of Flight*, no. 9, Smithsonian Institute Press, Washington, DC.

Millikan, Lt. Col. R. A., 1919. "Some scientific aspects of the meteorological work of the US Army." *Monthly Weather Review*, April 1919.

Mouginot et al., 2019. "Forty-six years of Greenland Ice Sheet mass balance from 1972 to 2018." *PNAS*, vol. 116, no. 19.

NOAA, 2013. "Service Assessment: Hurricane/Post Tropical Cyclone Sandy."

NTSB, 2017. "Report on the sinking of US cargo vessel *El Faro*, Oct. 1 2015," National Transportation Safety Board, Washington, DC, MAR-17/01 PB2018-100342.

O'Riordan, E., 2021. "US honor for 98-year-old woman whose Mayo weather report changed D-Day landing." *Irish Times*, June 20, 2021.

Pearce, R., 2005. "Why must hurricanes have eyes?" *Weather* (Royal Meteorological Society), vol. 60, no. 1.

Persson, A., 1998. "How do we understand the Coriolis force?" *BAMS*, vol. 79, no. 7.

Persson, A., 2006. "Hadley's Principle: Understanding and misunderstanding the trade winds." ICHM, vol. 3.

Revkin, A., 2015. "Blizzard questions, including why a European weather model (usually) excels at US forecasts." *New York Times*, January 26, 2015.

Rozwadowski, H., 2016. "Reconsidering Matthew Fontaine Maury." *International Journal of Maritime History*, vol. 28, no. 2.

Sandweiss et al., 2007. "Mid-Holocene climate and culture change in coastal Peru," in *Climate Change and Cultural Dynamics: A Global Perspective on Mid-Holocene Transitions* (Elsevier).

Schultz, D., and K. Browning, 2017. "What is a sting jet?" *Weather*, vol. 72, no. 3.

Schweiger, A., K. Wood, and J. Zhang, 2019. "Arctic Sea ice volume variability over 1901–2010: A model-based reconstruction." *Journal of Climate*, vol. 32, 4732.

Shapiro, M., and D. Keyser, 1990. "Fronts, jet streams and the tropopause." Erik Palmén Memorial Volume, American Meteorological Society, 167–191.

Sienkiewicz, J., and L. Chesneau, 2008. "Mariner's guide to the 500-millibar chart." *Mariner's Weather Log*, vol. 52, no. 3.

Smith, L., and S. Stephenson, 2013. "New Trans-Arctic shipping routes navigable by midcentury." *PNAS*, E1191–E1195, www.pnas.org/cgi/doi /10.1073/pnas.1214212110.

Staff Members of the Department of Meteorology of the University of Chicago, 1947. "On the general circulation of the atmosphere in middle latitudes." *BAMS*, vol. 28, no. 6.

Steffensen et al., 2008. "High-resolution Greenland ice core data show abrupt climate change happens in few years." *Science*, vol. 321, no. 5889, 680–684.

Stenhouse et al., 2014. "Meteorologists' views about global warming." *BAMS*, vol. 95, no. 7.

Struznik, E., 2019. "A Northwest Passage journey finds little ice and big changes." *Yale Environment 360*, https://e360.yale.edu/features /a-northwest-passage-journey-finds-little-ice-and-big-changes.

Sverdrup, H., and W. Munk, 1947. "Winds, sea, and swell: theory of relations for forecasting." University of California Scripps Institution of Oceanography, San Diego.

Transportation Safety Board of Canada, 2007. "Marine Investigation Report M06F0024: Crew member lost overboard, sail training vessel *Picton Castle*, 376 nm SSE of Lunenburg, Nova Scotia, 08 December 2006."

Transportation Safety Board of Canada, 2010. "Marine Investigation Report M10F0003: Knockdown and capsizing, sail training yacht *Concordia*, 300 miles SSE off Rio de Janeiro, Brazil, 17 February 2010."

United States Coast Guard, 2014. "Report 16732: Investigation into the circumstances surrounding the sinking of the US tall ship *Bounty*." US Department of Homeland Security, May 2014.

Wallace et al., 2014. "Global warming and winter weather." *Science*, vol. 343, no. 6172.

Walsh et al., 2018. "100 years of progress in polar meteorology." *Meteorological Monographs*, American Meteorological Society, vol. 59, no. 1.

Wood et al., 2018. "Results of the first arctic heat open science experiment." *BAMS*, vol. 99, no. 3.

Wood, K., and J. Overland, 2003. "Accounts from 19th-century Canadian Arctic explorers' logs reflect present climate conditions." *Eos*, vol. 84, no. 40, 410–412.

Wyrtki, K., 1975. "El Niño—the dynamic response of the equatorial Pacific Ocean to atmospheric forcing." *Journal of Physical Oceanography*, vol. 5, 572.

Yau et al., 2016. "Reconstructing the last interglacial at Summit, Greenland: Insights from GISP2." *PNAS*, vol. 113, no. 35.

Zebiak, E., and M. Cane, 1986. "A model El Niño-southern oscillation." *Monthly Weather Review*, vol. 115, 2262.

ACKNOWLEDGMENTS

I will begin by thanking the people who convinced me that this book was worth writing, and that I was the one who should write it. First among them is Dan Brayton of Middlebury College, the world's toughest English professor and a shipmate of many years' standing. Also Rich King, who shared countless hours with me in mid-ocean, talking about how it all might look if put on a page. Amanda Moon was of tremendous help when the time came to turn my loose bundle of ideas into a book proposal, while my agent, Zoë Pagnamenta, deserves substantial credit for helping me determine which parts of this story were the most worth telling, as well as her hard work in finding a home for the project. Sincere thanks to editor Brent Howard and his crew at Dutton, whose able helmsmanship steered me along on the actual course of writing, and to Matthew Twombley, who applied his considerable talent to create images for the book that were both engaging and accurate.

ACKNOWLEDGMENTS

I am especially grateful to Captain Andy Chase, a standout among my many mentors, and Professor Kirk Maasch at the University of Maine, who introduced me to weather as an academic discipline, woven through with unlimited subparts to teach and learn. Thank you to my parents, Steve and Sandra—both writers and sometime sailors—who raised their children in a house filled with books and a resolute intolerance for bad grammar. Wandering off to sea seemed a fine thing to them; dangling modifiers surely not.

This is above all a book about the shipmates whose talents and energy have made my own career possible. There is no page long enough to thank them all, but I am particularly grateful for my time with the crews and staff of the SEA Education Association of Woods Hole, Massachusetts—a unique organization where what I learned goes well beyond the simple measure of days at sea.

Thanks finally to Karen, who knows many things.

ABOUT THE AUTHOR

Elliot Rappaport has sailed as a captain in the US maritime industry since 1992, involved primarily in the training of other mariners aboard a variety of traditional sailing ships. Presently a faculty member at Maine Maritime Academy, preparing cadets for professional careers at sea, he has also worked extensively at the SEA Education Association in Woods Hole, Massachusetts, an organization that offers shipboard programs in ocean science and leadership to college undergraduates. A graduate of Oberlin College and the University of Maine, Elliot lives in coastal Maine when not at sea.